Discovering America

Discovering America

JOURNEYS IN SEARCH OF THE NEW WORLD

Timothy Jacobson

BLANDFORD

Page ii:
The rugged coast of eastern North America must have seemed desolate and forbidding to early European mariners. This stark outline of Labrador, unchanged since the time of the Norsemen who first ventured across the North Atlantic, was the threshold of the New World.
(Arnold Zageris)

First published in the UK by Blandford, an imprint of Cassell, Villiers House, 41-47 Strand, London WC2N 5JE

Distributed in Australia by Capricorn Link (Australia) Pty Ltd, PO Box 665, Lane Cove, NSW 2066

Copyright © 1991 Key Porter Books Limited

Published in Canada by Key Porter Books Limited, 70 The Esplanade, Toronto, Ontario M5E 1R2

British Library Cataloguing in Publication Data
Jacobson, Tim
　　Discovering America.
　　1. United States. Visitor's guides
　　I. Title
　　917.304927

　　ISBN 0-7137-2245-2

Typesetting: Images 'N Type Ltd.
Printed and bound in Hong Kong

Contents

Europe Moving Out

TO MAKE KNOWN OR VISIBLE, TO OBTAIN SIGHT OR knowledge of for the first time, is to discover. And this is what happened when seafaring Europeans set out across the world's great oceans in the fifteenth and early sixteenth centuries. They came upon things that neither they nor any men like them had ever seen before, *and* they lived to get back home and tell the tale. Some of those sights were of Stone Age peoples whose favor could at first be bought for a few cheap trinkets (as was the case in most of America north of Mexico). Sometimes they were of cultures whose wealth and sophistication put Renaissance Europe to shame (as was the case when Europeans first wandered into the interior of the Indian subcontinent in the sixteenth century). A highly ethnocentric lot, as scholars would say today, these Renaissance Europeans packed in their psychological and intellectual kit an arrogance that brooked no competition and left in its wake no shadow of a doubt that the world was indeed theirs to discover. Their point of view was exclusive; their power and their will to subdue were in time irresistible.

The answer to the question of just who discovered whom, when the Christian white man waded ashore armor-clad with cross and arquebuses close at hand to meet the savage red man, depends on one's point of view. Perhaps the most value-neutral way to describe the events of discovery is not to use the word "discovery" at all but rather to say that the isolation of whole civilizations (up to this time not that unusual) was at last coming to an end. Very soon there simply would be no more "lost worlds" to find. But from the perspective of half a millennium, "discovery" seems all the more fitting a description for what happened, its object the men who watched from the bushes with amazement, fear and loathing, its subject the people in the strange ships who had a plan in hand and the morale and technology to implement it.

A replica of the Golden Hinde *commemorates Francis Drake's 1577 circumnavigation of the globe. For Europeans, discovery of a vast unknown land across the Atlantic heralded an age of promise and adventure. (Jürgen Vogt/The Image Bank)*

These men, our subjects, were the products of the Europe they came from, which for decades they would still think of as home. Like their vastly more numerous fellows who never left that homeland, they embodied both the ingrained convictions of the Middle Ages and the newer attitudes associated with the Renaissance. The Renaissance in Europe coincided with the voyages of the great discoverers: Machiavelli and Michelangelo and Leonardo wrote and painted while Cabot and Cartier and Raleigh sailed and adventured and sailed home again. So it was with the Reformation. Luther, Cranmer and Calvin reformulated the received faith of their fathers just as cartographers reconfigured Ptolemy's classical picture of the world and applied new theories to anticipate what might likely be found out there beyond the range of the known world. The humanist impulse of the Renaissance, which reimagined man as a self-confident, self-respecting and self-conscious creature, was no doubt fiercely embodied in the discoverers, pirates and plunderers who first forced Europe's ways onto America. And the competitive impulse of the Reformation, which released the contentiousness of militant believers to battle over speculative matters of the spirit, no doubt added even more fuel to the striving of discoverers utterly convinced of the righteousness of their mission.

But by the time of the great American discoveries, neither Renaissance nor Reformation had anywhere near washed clean the slate of feudal Europe, where was recorded a thousand years of habit. The humanism of Erasmus raised strange and wonderful new prospects, but it did not, either in him or in any of his contemporaries, introduce skepticism enough to subvert the pieties of a still profoundly religious age. Luther's reforms broke finally the universal reach if not the universal claims of the Catholic Church and injected Christianity with a freshness and urgency it had not known since before the fall of Rome, but they did not break or intend to break with the medieval notion that, whatever it was, there could be but one true faith. The Reformation introduced competition (frequently barbarous competition), not toleration, into the religious affairs of men, at a time when things secular and things religious had just begun to diverge. Further in the future yet lay the dawn of political revolution in Europe (the nation-states themselves in the fifteenth century having just been consolidated in powerful monarchies). No significant strain of Renaissance political thought anticipated anything resembling familiar modern forms of participatory government, or ever seriously imagined anything but a deferential hierarchical political order in which the few ruled the many and rightly so.

When thinking about the changes that flowed from the Renaissance and about the coming of new ways of looking at man and his world, it is important not to dismiss the old Europe of the late Middle Ages as a period beset by backwardness and willful obscurantism. Europe, as the Renaissance and the Reformation were waiting to happen, was a very busy and sophisticated place. Most

The Flemish carrack was built for trade in Europe's coastal waters and narrow seas, but the design evolved into the ships that became the familiar vehicles of discovery. (Science Museum, London)

people of course still lived and died within one small orbit, their baptisms and burials recorded in one parish book, and so it would remain until the wrench of the industrial revolution and the growth of cities in the late eighteenth and nineteenth centuries supplied enough energy to fracture forever the fixed patterns of a predominantly rural world. But, even in that old world, commerce over great distance abounded, and travel, for those who needed to, was practiced and practical. Artists, ambassadors, craftsmen, clerics, soldiers and merchants moved regularly about Europe despite what would seem to us appalling physical obstacles.

Just as important, and as typical, Europeans moved beyond Europe, to Africa and the Near East, to Russia and India and the Orient. They did so laboriously, slowly overland and hopscotching along far coasts and through inland seas. Venetian and Genoese ships linked the Mediterranean world with the spice-rich East via Constantinople, and caravan routes stretched across Central Asia as Europeans in the name of God and commerce sought souls and profits among the Mongols. Indeed, for sheer derring-do and vast distances covered, little in the history of exploratory travel can match the history of the hundred years between 1245 and 1345.

With the death of the Great Khan in 1241, the Mongol war machine that had terrified Christendom and carved out an empire

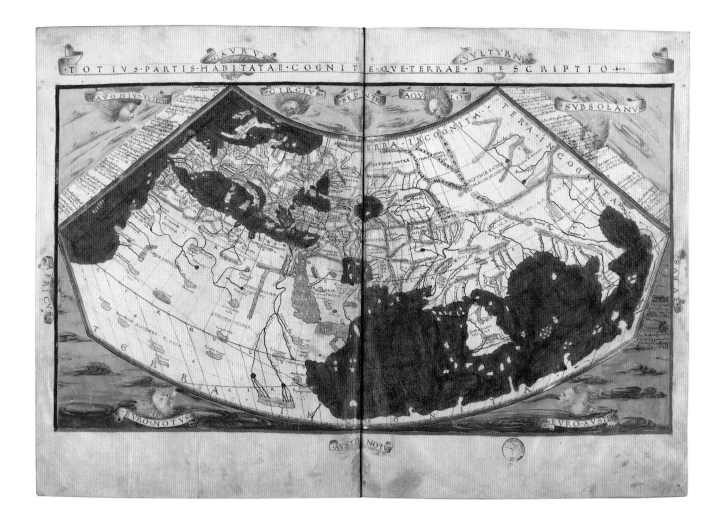

Second-century geographer Claudius Ptolemy left a legacy that lasted a thousand years. His rendering of an Asia that reached much farther around the globe than it in fact does made reaching it by sea seem possible. What he missed, and what the discoverers subsequently stumbled onto, was America. (Biblioteca Apostolica Vaticana)

reaching from the Sea of Japan to the edges of Europe paused and then redirected itself southwestward against the lands of Islam and the caliphates of Baghdad and Syria. In this providential respite, Europe saw twin opportunities of alliance with the Mongols against Islam, the common enemy, and of converting to Christianity the Mongols themselves, who were known to be a not especially religious people. Thus began the epic journeys of the Franciscan friars: the Italian Giovanni da Pian del Carpini, who traveled as a papal ambassador to the court of the Mongols at Karakorum in 1246 and wrote a book about it (*Historia Mongolorum*), and the Flemish William of Rubruck, who as a missionary made the journey into Mongolia in the early 1250s. If the Mongols did not quickly flock to cross and altar, they did see practical value in contact with the West, and under their most enlightened chief Kublai Khan, who ruled from 1259 to 1294, a sort of *Pax Mongolica* opened Eurasia, from the Volga to the Great Wall, to the possibility of trade with Europe.

Merchants followed the missionaries, the most legendary from the Polo family of Venice: brothers Niccolò and Maffeo, and Niccolò's son Marco, who penetrated all the way to Peking and Xanadu, Kublai Khan's fabulous summer residence to the north. Marco stayed on in China in service to the Mongols for sixteen years, compiling the mass of information by which Westerners of the next century chiefly came to know the Far East and which has made Marco's name something of a household word ever since.

Along the caravan routes the traders for the next fifty years exchanged their goods directly for the exotica of Asia without the necessity of Levantine middlemen. The commerce lasted only as long as the religion of the unreligious Mongols still lay in the balance, and in the end distance and culture prevailed against the missionaries' heroic efforts to convert the khans and bring their hordes into the fold of St. Peter. Rather, it was Islam that prevailed in the strategic khanates of central Asia where Europeans had briefly enjoyed such peaceful and profitable passage along the "Golden Road to Samarkand," and a resurgent Islam promptly closed the overland route to the Christian infidel. The eastern portals of Asia shrank further with the advance of Islamic Turks across the Near East, culminating in the fall of the great Byzantine entrepôt, Constantinople, in 1453. The curtain that descended between East and West was hardly an iron one, but the obstacles imposed by the Turks on the land routes through Asia Minor and the sea-lanes across the eastern Mediterranean and east of Suez made commerce ever slower and more perilous—and therefore ever more expensive.

The key commodity whose supply these developments radically reduced was spices—pepper, cloves, nutmeg, cinnamon and an array of other aromatic condiments. And it happened just as European demand for spices was reaching new highs. Spices made palatable the taste of bad food, enhanced the enjoyment of good food and were used as preservatives. Spices found their way into the pharmacopoeia and generally took on all the desirability that fashion could assure. Their chief source was the storied Spice Islands of the Moluccan Archipelago between the Indian and Pacific oceans,

In Juan de la Cosa's world map, spice caravans across central Asia are depicted with greater confidence than any sea routes westward into the Atlantic. (New York Public Library, Astor, Lenox and Tilden Foundations)

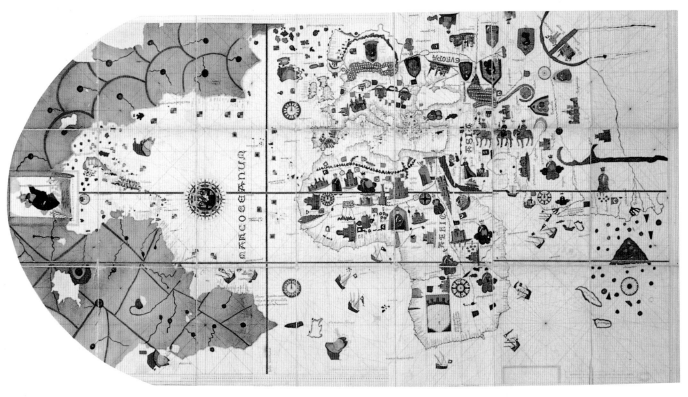

but East Asia in general supplied them, along with silks, drugs and ceramics. While Christian trading states like Venice kept a modicum of trade moving and while smugglers strove as ever to fill the void, Europeans at the end of the fifteenth century found that the market offered less and less of what they craved more and more to buy. The situation was aggravated by the scarcity in Europe of precious metals, especially gold, which came from sub-Saharan Africa and was subject to tribal instability and profiteering Arab middlemen. Even as hoarding increased, accelerating economic activity created ever greater demand for these metals.

One thing the West did not want from the East, but got anyway, was the plague or Black Death, which was carried from China on the caravans to the Crimea, then to the Mediterranean and north in the 1340s. It killed millions horribly (a third of the population of England alone) as far north as Scotland and Scandinavia, and while it cast a sudden temporary pall over the economic life of the whole continent, it may also have served as a tonic. By reducing the supply of labor, the plague helped hasten the decline of feudal and manorial institutions and so make way for the rise in the next three centuries of unitary nation-states with the ambition and the wherewithal to look outward. Banking and finance would re-emerge in its wake refreshed and in time would take on the mercantilist and then capitalist character that helped propel overseas discovery and settlement. Western Europe in the fifteenth century slowly began to assume the shape it would have right up to the French Revolution: powerful and mutually suspicious England, France, Spain and Portugal, and small but shrewd Holland. Each took its place in the first burst of overseas expansion that carried European ways and dominion to the true ends of the earth.

From their perch on Europe's southwestern promontory, the Portuguese made the first tentative thrusts out, but in an eastward, not a westward, direction. East, after all, was where the spices came from, and what better way to outflank the Turks and Arabs than by finding an all-water route to the Indies? Between the 1430s and the 1480s, the captains of Prince Henry the Navigator probed gingerly south and west from the Canary Islands along the African coast in the hope that Africa could be traveled around on the bottom end and the ocean path east made plain. Slowly their stone marker pillars went up: Cape Bojador, 1432; Cape Verde, 1444; the Congo River, 1482 (whose silt discolors the ocean 300 miles out from shore); Cape Cross, 1485. And finally, in 1488 Bartholomeu Dias doubled the Cape of Good Hope and entered the Indian Ocean. Ten years later, Vasco da Gama pressed beyond, up the east coast of Africa, and then struck out for India, which with the help of a local pilot he reached in less than a month. (The miserable return voyage, beating against the monsoons, took four months.) The Portuguese then bluffed and blasted their way to a vast East Indian empire that made them rich beyond wonder, from Goa and Cochin to Ceylon (the world's finest source of cinnamon), to Malacca, Java, Timor, the Moluccas, Canton and as far as Tanega Shima in Japan.

The Spanish, their Iberian neighbors, were busy in the other direction, sailing west to get east, not to prove the world was round (or elliptical), which was then a widely accepted proposition, but to find yet a shorter route to the same spice-rich East. That it might actually be shorter to go the other way around was not at all an unreasonable speculation, given the erroneous maps of the time, which vastly overestimated the length of the Asian landmass and thus underestimated the width of the ocean waste lying in the way (while of course never showing that another whole continent— the Americas—also barred the way). By the end of the fifteenth century it looked to Europeans, with already much experience in blue-water ocean sailing to the Canaries, Madeira and the Azores (which are fully a third of the way across the Atlantic), like a doable thing. It is fortunate that their information was bad, for had the real distances been known, likely no one would have been fool-hardy enough to give it a try.

On his first voyage west in 1492, Christopher Columbus, a Genoese sailing for the king and queen of Spain, first stumbled across the awkward American obstacle (or a small outpost of it): Watlings Island in the Bahamas, which he called San Salvador. Naked natives greeted him, and he called them Indians, dead certain that he had found the realm of the Great Khan. On his fourth and final voyage between 1502 and 1504, Columbus at last sighted the mainland, while coasting along the American shore from

The landing of Christopher Columbus on the Bahamian island, which he named San Salvador, has been the subject of more renderings than any other event of discovery. This engraving offers an especially fulsome panoply of characters, and surely exaggerates what actually took place on October 12, 1492. (National Archives of Canada)

Honduras to Panama and reporting that without a doubt the River Ganges was but a ten-day march away. The great mariner went to his grave in 1506 convinced as ever that Cuba was not Cuba but China.

However imprecise Columbus's understanding of where he actually was, his avaricious patrons were anything but slow in taking advantage of his claim in their name. From it followed within thirty years Spain's Central and South American empires, the abrupt end of the Aztec and Inca civilizations and, for the royal treasury in Madrid, a king's ransom in gold, silver and precious stones. It was even better than the "real" Indies, and in a gold-hungry Europe Spain's good fortune in hitting the jackpot radically raised the stakes when the rest of Europe got ready to seek out its own share. But the luck of the Spaniards was not to be repeated: no discovery made by England or France in North America ever yielded half the booty that both the English and the French, with some regularity, looted from the Spanish treasure fleets.

Visions of sea monsters revealed both the power of Europeans' imagination and the extent of their ignorance when it came to the sea that separated them from new worlds. Whales and walruses were, in fact, the most monstrous sea creatures that nature could offer. (John Carter Brown Library, Brown University)

If the Europeans had the motives for discovery, they also had the means. They were means that had been evolving over centuries of close contact with the sea, and the job now at hand was first of all a seafarer's job. Two shipbuilding traditions came together in the small vessels that made the American discovery. From the Mediterranean came lateen-rigged coasting vessels that for centuries had plied the trade routes from the Iberian Peninsula, Genoa and Venice to the Levant, and which the Portuguese and the Spanish had adapted for their long, probing voyages down the African coast in the fifteenth century.

From the north of Europe came the sturdy and stout square-rigged merchantmen descended from the Scandinavian *knarr*. Not the famous oar-propelled longships of the Viking raiders, the *knarrs* were workaday freighters, open boats best suited for voyages in sheltered European waters. They did, however, make the stepping-stone crossing of the North Atlantic via Iceland and Greenland to Newfoundland and Vinland five hundred years before Cabot and Columbus made the crossing direct. In the Hanse ports of Germany and in the Low Countries, the *knarr* evolved into the *cog* or *cogge*, which was the true prototype of the discovery vessels. Completely decked over, with a true rudder hung to a sternpost and controlled by a hardwood tiller, the *cog* was a match for the fiercest seas. Clinker-built hull construction (where the planking overlapped board on board) gave way to flush carvel-built hull work, which when well-caulked with oakum (shredded hemp fibers) and pine tar pitch made a tight ship for ocean sailing. Masts and rigging and sail-plans were adapted to meet new needs. By 1500 three masts were the rule (oars having faded away with improved techniques of sailing to windward), mainmast and fore-mast carrying square courses made of linen, which were relied on to provide the drive. A triangular lateen sail hung from the

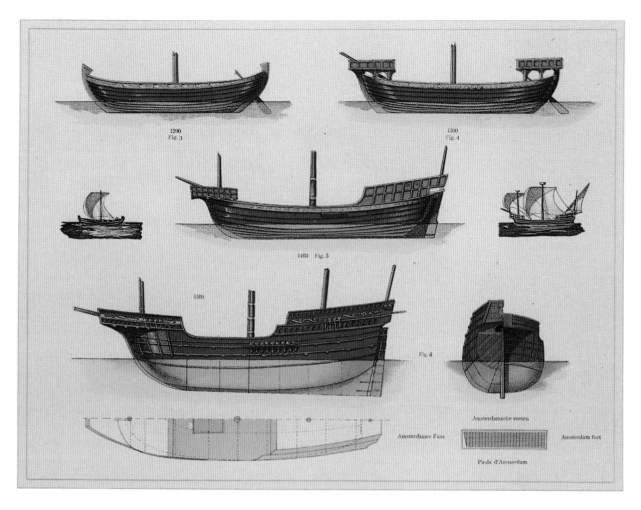

The evolution of European sailing vessels from the late Middle Ages to the eve of discovery reflects a growing capacity to sail oceanic distances—and sail home again. (Thomas Fisher Rare Book Library, University of Toronto)

mizzenmast and provided help in tacking and sailing on the wind. By 1600 it was also common to see square spritsails on the bowsprit and larger topsails on mainmast and mizzenmasts, which made for smarter maneuvering with smaller crews.

Two decks were standard, one just a few feet over the ballast, the top or spar deck above it exposed to the elements. An aft superstructure or sterncastle housed steering gear and the master's cabin with chart lockers, a traverse board and the half-hour glass, then a ship's only timepiece. Forward, a forecastle, or fo'c'sle as it came to be called, was home to the ordinary seamen. In between lay the low waist of the ship, closest to the waterline, where the small boats were stowed (though a longboat or pinnace frequently had to be towed behind, or carried in pieces in the hold to be reassembled when needed for work in shoals and shallows). The ship's hold, a cavernous and usually undivided space below, carried the cargo: freight in barrels and chests, or, as time went on, passengers bound one-way for New World settlements. It was the capacity of a ship's hold to carry double-hogsheads of wine, or tuns, that measured its tunnage, or the burden it could bear. The modern maritime ton of some forty cubic feet derives from this early usage. Such ships developed the Newfoundland fishery, sailed regularly to Iceland, the Azores and eventually around the Cape to India and the Spice Isles. They were not as a rule large vessels built in

Mapping the New World

THE NOTION THAT THE EARTH WAS ROUND WAS NOT A new one to the age of discovery; what was new was the discovery of its great size—and how much of it was water and not land. The maps of antiquity, particularly those of the second century geographer Claudius Ptolemy, left a legacy that lasted a thousand years. They portrayed a round planet and detailed knowledge of the Mediterranean world. Beyond that, however, the image grew fanciful and, as it turned out, was almost always mistaken. Ptolemy pictured Asia as being vastly larger than it actually is, saw the bottom of Africa wrapping around into Terra Incognita at the bottom of the world, and the Indian Ocean consequently as a vast enclosed lake. No one had guessed that the Pacific, the greatest of the seas, was even there. Nothing but the experience of discovery and the careful observations of the discoverers changed this.

As a result of Columbus's voyages and Portuguese ventures eastward, maps such as the Cantino Map made in 1502 revealed remarkable new knowledge: the coastline of Africa is shown with great clarity, and Africa has been completely cut loose from Terra Incognita, thus turning the Indian Ocean into a proper sea. The West Indies appear as meticulously reported by Columbus, and the coast of South America as far as Brazil takes on a reasonable likeness to reality. There is even a hint of Florida and the North American mainland. By the time Magellan's ships first circumnavigated the world in the 1520s, maps routinely filled in the eastern coastline of the New World all the way from Labrador to the Strait of Magellan, and indicated the vast expanse of the Pacific. Still missing was any notion of the size and shape of the American continents (just as the interior of Africa was still largely a blank space) or any indication of islands in the Pacific and the shape of Japan and the mainland of East Asia.

French cartographer Pierre Desceliers's 1546 world map incorporates information from Jacques Cartier's three voyages to the New World. (John Rylands University Library, Manchester)

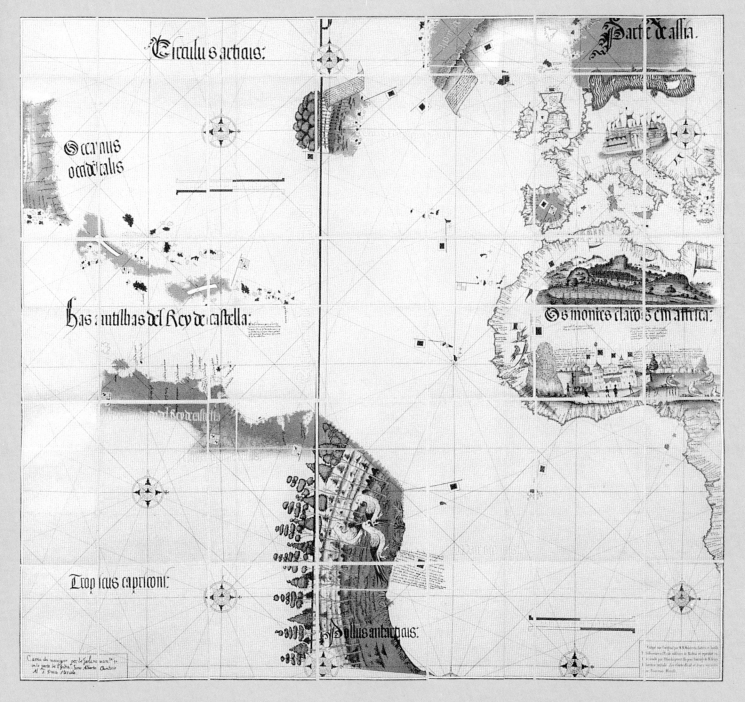

For mariners the problem with all maps (even as the empty places got filled in) was the difficulty of representing a round world on a flat chart. The difficulty of charting an east-west course through unknown waters increased the farther one sailed from the equator as lines of longitude converged, and it was not until 1569 that the Flemish geographer Gerhardus Mercator worked out a successful round projection on a flat surface by distorting almost beyond recognition the far northern and southern regions and keeping the zones that had most of the shipping traffic in the correct proportion. Smaller, separate polar projections were added at the corners of the chart. For centuries men had charted the heavens while they knew little of the earth, but in the age of discovery star charts improved along with maps of the seas and continents as ships sailed south of the equator and brought back a picture of the southern skies.

The Cantino Map of 1502, just ten years after Columbus's first voyage, showed the West Indies, the coast of South America as far as Brazil and, in the north, even a hint of Florida on the North American mainland. (The Newberry Library, Chicago)

naval yards under official auspices, but rather were local products fitted to the knockabout needs of trade.

The tunnage of none of the ships we encounter in the American discovery was very great, and illustrated histories of the subject are fond of emphasizing their puniness compared with ocean liners and supertankers. (One makes the point by superimposing Sir Francis Drake's *Golden Hinde* crosswise on a tanker, its length not even equal to the width of the modern behemoth.) John Cabot's *Mathew*, which sailed from Bristol to Newfoundland in 1497, was but fifty tons burden; *Susan Constant*, the largest of the three ships that sailed to Jamestown in Virginia 110 years later, was just 120 tons. But seaworthiness and suitability for both open-ocean sailing and coastal exploration are not a function of great size. Vastly larger ships had for centuries plied the Mediterranean, and the mariners who took these cockleshell vessels across the Atlantic knew exactly what they were about, and had coolly calculated the odds that such ships not only could sail out across vast open seas to some far shore but could also safely bring them home again. They were not much different in size from many modern sailing yachts that have made the Atlantic passage for fun and recreation. But when the stakes and the risks were higher, there were, in the learned view of the great sailor-historian Samuel Eliot Morison, no better vessels imaginable for the task at hand than these graceless bundles of oak, linen and hemp.

Their successful navigation depended on a few simple instruments and on a seaman's feel for his ship and the sea. If one did not venture too far from a coastline, then compass for headings, lead line for soundings, the log to determine speed (not invented until the late sixteenth century), and the lookout for hazards were

On Desceliers's world map of 1550, Africa and the Indian Ocean remain the dominant focus. Over the next century, that focus would shift to the West. (Courtesy of the Trustees of the British Library)

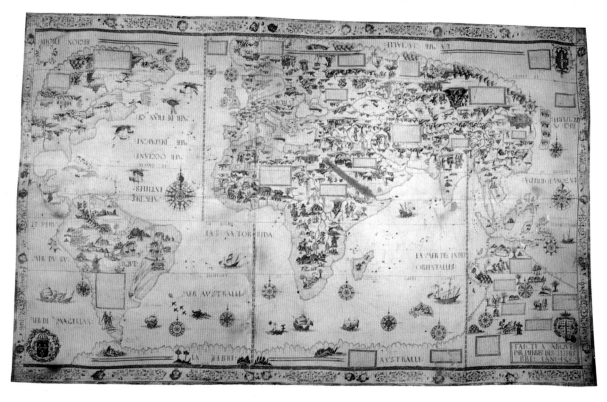

Carracks, such as those in this 1521 rendering, extended the empire of Portugal eastward from the Cape to the China coast. But with the exception of Brazil, empire in the West went to other powers. (National Maritime Museum, Greenwich)

enough to steer by. Transoceanic voyages required, even to fix approximate positions, use of celestial navigation. By the late fifteenth century, the best mariners (who were a minority of all mariners) were using the nonreflecting quadrant and astrolabe to take the altitudes of heavenly bodies—which meant the North Star (Polaris), its "guards" (the two outer stars of the Little Dipper), and the sun—and declination tables were available to determine latitude from the meridional observation of the sun. The crucial capacity to determine longitude, and thus truly to "fix" a position on the earth's curved surface, awaited development in the eighteenth century of the chronometer and the ability accurately to tell time. In the age of discovery, shipboard time was measured only by the sands of the ship's glass, which had to be turned every half hour and was vital to navigation by time, distance and direction—the "dead reckoning" for which many of these early mariners had an uncanny sixth sense. (Every eight glasses—four hours—another watch was called on deck, an event later signaled by the ringing of a ship's bells.)

Given the simple navigational tools available to them, the discoverers made blue-water voyages remarkable for their speed and directness; they knew how to make good use of the Trade Winds outbound and the Westerlies at higher latitudes heading home. Trouble usually came close to land, where tides, shifting winds

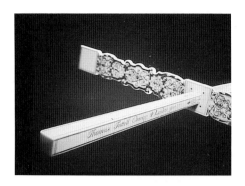

The cross-staff determined a ship's position north and south relative to the equator: a simple stick fitted at right angles with sliding crosspieces which, when aligned with the sun at one end and the horizon at the other, yielded the sun's altitude. A 1669 engraving shows the cross-staff in use. (National Maritime Museum, Greenwich)

and currents always threatened a quick and nasty end to the most serene of ocean sails. At night the coastlines of the world were as yet largely unlit—the brightest lights anywhere then were dim by modern standards, lighthouses along the French and English shores few and far between and frequently illumined only during storms. The American coast was of course the darkest of all. Thus, near land, the ocean bottom frequently was the best guide to a safe haven, and skillful use of the lead line (which not only measured depth but was armed with tallow to bring up samples of the sea bottom) enabled a knowing pilot to read the changing nature of what lay beneath his keel and so literally feel his way ashore. Every stop on the jetty-less American shore also meant, on these early voyages, casting anchor, and ships usually carried at least four of the great iron hooks, which eventually became the very symbol of safety and seamanship itself.

The maps and charts that guided these sailors today seem fanciful, but then they were plausible products of a world still fixed on visions of Cathay and cities of gold beyond the western sea. Three sorts of maps helped men envision the way the earth's surface might be laid out. A product of the Middle Ages, the so-called Jerusalem maps ordered all creation around that holy city. With East, not North, commonly at the top, they were more memorials to the faith than guides to travel. Ptolemaic world maps (named after the second century A.D. cartographer Claudius Ptolemy of Alexandria) portrayed a more realistic world, at least as it was known in classical times. Europe, North Africa and the Near East came off in fair likeness to reality. The Indian Ocean was vast, but cut off from the Atlantic by an Africa that was joined along the bottom of the world to the mythical Terra Incognita. In the great expanses ascribed to Asia was left room to inscribe all the information brought back along the caravan routes by Marco Polo and other medieval travelers. By the end of the Middle Ages, it was widely believed that the world was round. Strabo, a classical writer who preceded Ptolemy, had said that the world formed a circle, itself meeting itself, which meant (if the Atlantic were not impassably wide) that the East could be reached by sailing west. So if you sailed off one side of a Ptolemaic map, it followed that you would sail back onto the other. Just how far it was around the back side remained a bit of a mystery, and the existences of yet another ocean (the Pacific) and yet another continent (America) were the final and greatest surprises.

By the end of the fifteenth century, mapmakers were getting the theory right (though there would be some bad and costly mistakes still ahead: about the existence of a Northwest Passage, for example), and it was a theory vital to mariners and backers risking much on ventures across seas where no one had sailed before and returned to tell about it. The third type of map, the portolan, reflected practical techniques of spacing the features on a map, such as capes, bays and islands, according to the actual compass bearings one needed to plot a real course between two points.

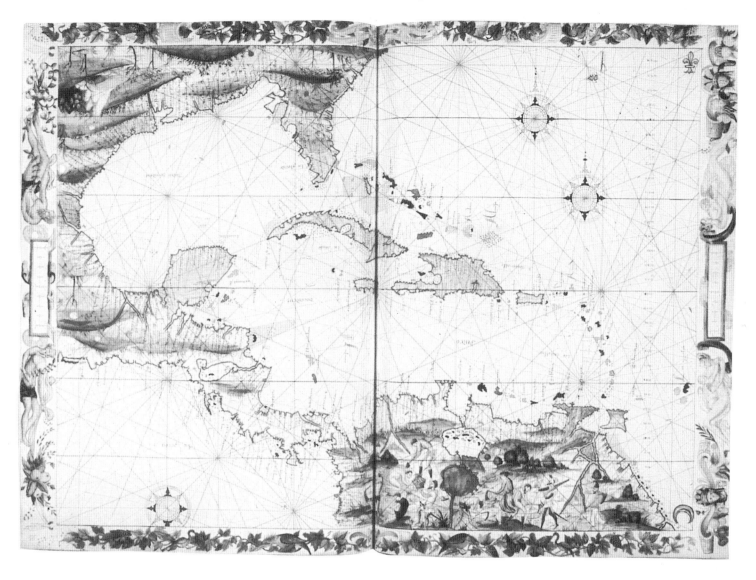

These were local and regional maps that recorded facts already known, and were meant for use by real mariners. As the first surveys filled in details of the American coast, portolan-like maps of the New World, complete with artistic renderings of what it was imagined the interior spaces might contain, joined those of the old.

The naming of that new world is commonly ascribed to the German mapmaker and geographer Martin Waldseemüller, whose world map of 1507 portrayed all the reported western discoveries from Labrador to the Argentine as part of one contiguous landmass. He called it America after Amerigo Vespucci, who in 1501 had coasted from Guiana to Patagonia and established the great length of South America. In 1519, Ferdinand Magellan sailed around the southern tip of South America (actually through a strait very near the tip, which ever since has borne his name) on a voyage that would lead around the world. En route, he at last found the Spice Islands far westward across a vast Pacific Ocean that conquistador Vasco Nuñez de Balboa had set eyes on from the Isthmus of Panama in 1513, and that Waldseemüller also had shown on his map. But it remained for Giovanni da Verrazzano in 1524 to fill the last gap, sailing northward from the edge of Spanish

The Caribbean lay at the heart of Columbus's discoveries, and for three centuries defined the dangerous conjunction of Spanish and English interests in the New World. (Koninklijke Bibliotheeck, The Hague)

territory in Florida to Nova Scotia, to confirm that in the absence of the fervently hoped for passage through the barrier, America must be one continent. Verrazzano never guessed at its breadth and indeed thought that he had spied the Pacific across the narrow sand barrier of North Carolina's Outer Banks. (What he saw was Pamlico Sound.) "My intention in this navigation was to reach Cathay, and the extreme east of Asia," he confessed, "not expecting to find such an obstacle of new land as I found." Belief in a Northwest Passage would persist for years to come, but the search for it only confirmed the continental dimensions of the place the discoverers had stumbled across, and brought home the understanding that America was not a place to circumvent, but to settle.

Settlement when it came, however, would have eluded the most willful Europeans had they not possessed the weapons needed to make discovery stick once they had made their voyage and got ashore. The armaments at their disposal from the Iron Age alone— their armor, pikes and swords—with which every discovery expedition was well provided, were superior to what the most advanced native American peoples, the Aztec and the Inca, could muster. The Europeans were practiced with the longbow and the mechanized crossbow, and of course with the new percussive firearms that would dominate the weaponry of warfare until our own time. Their ships and fortifications mounted cannon; their soldiers shouldered arquebuses and packed pistols.

While the attitude of the indigenous peoples whom the Europeans confronted in North America was frequently at first only curious, and changed to hostility only when it became clear that the strange visitors had come to stay, the Indians lacked utterly the military resources, to say nothing of the political discipline, to turn back so advanced a foe. Had the Europeans found themselves in America faced off against adversaries who had reached comparable levels of arms technology and military organization (as they did in China), then dominion over North America would have been even slower in coming than it actually was.

For we must recall that fully 110 years elapsed between John Cabot's voyage to Newfoundland in 1497 and the establishment of the first English settlement that came to be permanent, at Jamestown in 1607. During those long years, the energies of England and France, the two kingdoms who would have the greatest role in settling North America, lacked the fierce concentration that Spain was able to bring to her conquest of the Americas south of Florida hot on the heels of Columbus's voyages in the 1490s. Had the English and the French, at this early tentative moment, met determined resistance by powerful native peoples, there is nothing to say that not all their armor and artillery could have prevailed to subdue the New World wilderness. But they did not meet such resistance. By the time they determined to possess what it was they had discovered, in the late sixteenth and early seventeenth centuries, their intentions fairly well matched their

Giovanni de Verrazzano was born near Florence in 1485, but sailed under the banner of Francis I of France in 1523 and 1524 to explore the American coastline north of Spain's possessions. He probed from Cape Fear on the Outer Banks of North Carolina northeastward to Nova Scotia, discovering as he went New York harbor and Narragansett Bay, and concluded that this coast belonged to a single continental landmass. (National Archives of Canada)

In Jean Rotz's map of North America and the West Indies, the coastlines bear some general resemblance to reality, but interiors remain utterly unknown. (Courtesy of the Trustees of the British Library)

abilities: curiosity, commerce and conquest lured them on until eventually there could be no going back.

It is important to separate first intentions from final results. Willy-nilly, one thing always leads to another, and so here too. From the first discoveries of the American surprise and from the early efforts at settlement, great new nations sprang up. The results, in the United States and Canada, were free and prosperous and stable states whose model in the world is much admired. On national holidays, it is common to celebrate origins as well as accomplishments, and the origins in the case of the discovery of America are well adapted to the exercise. For except to perhaps the most militant partisans of the departed Indians, the history of discovery seems such an epic tale, such a chronicle of large deeds done by truly brave men. Moderns may recoil at some of their brashness, their utter self-absorption, their aggressive patriotism, their religious intolerance, but we cannot help but admire the supreme cultural confidence at the root of these things.

But awe felt from afar—five hundred years—for such a remarkable cast of characters acting out their drama on open oceans and wild new continents should not cause us to forget how far "afar" really is, and how different the world view of the early explorers was from ours today.

The Vikings saw the first glimpse of America that we know about. Driven by pressures of population in their native Norway, the Norsemen made progressive steps westward across the North Atlantic, first to Iceland and then to Greenland. Vinland, as the Icelandic sagas three centuries later referred to it, was yet another land farther to the west, some part of it probably on the American mainland or perhaps Newfoundland. The Norsemen came across it in the fog while looking for Greenland: they had not planned their discovery. Yet in the last years of the tenth century they made a brief settlement there. We do not know if they intended to stay, only that they didn't, apparently succumbing to the weather and inhospitable natives. By the time John Cabot came to Newfoundland five hundred years later, even the Norse settlements on Greenland had been given up. The whole episode is shrouded in mystery still. Archaeological evidence definitely confirms their presence in northern Newfoundland, though it does not tell us whether Newfoundland was in fact Vinland. But wherever Vinland was didn't really matter to the medieval Europeans because Europe was not yet ready for the news.

By the time of John Cabot in the 1490s, it most certainly was. Cabot set out twice for Henry VII of England. He came back only once, but once was enough to stake England's claim. Probably a Genoese, most certainly an Italian, Cabot remains a mysterious figure: no portrait of him exists and we have no sea journal that he kept. An experienced mariner with an ambition to find a shorter route to the spice-rich East, Cabot may have been seeking support in Spain in 1493 when Columbus returned triumphantly with his news of a discovery to the west (which he said was really the East). It must have been a depressing moment, and it drove Cabot to England, whose king, the founding Tudor, Henry VII, had earlier refused support to Columbus and no doubt now regretted it. This time Henry jumped, at least to the extent of granting Cabot letters-patent.

Cabot sailed down the Avon from Bristol with but a single ship, the *Mathew* (Henry granted favor but no up-front money), and thirty-three days later found, or found again, Newfoundland, landing on Quirpon Island at the tip of the Great Northern Peninsula and, unbeknownst to him, only a stone's throw away from the landing site of the Norsemen half a millennium before.

In the 1530s and 1540s, Saint-Malo master pilot Jacques Cartier explored the mainland of what would become Canada, discovering one of the continent's two greatest rivers, the St. Lawrence, and fixing the claims of France to a large portion of North America. But when Cartier paid his visits to this shore, most on his mind was not Canada but the elusive old short route through it to the Indies. He spent a horrific winter at the site of Quebec City and probed the St. Lawrence as far as the Lachine Rapids above Montreal before turning back. He probably set a record for the

This sixteenth-century map of the North Atlantic, drawn in Iceland, shows medieval Norse discoveries and settlements. (Det Kongelige Bibliothek, Copenhagen)

amount of iron pyrites (fool's gold) that could be packed into one ship, and back in France he traded till his death on the belief that Canada and the treasures of the legendary Kingdom of Saguenay could do for France what the wealth of the Aztec and Inca had done for Spain. That he was mistaken does not lessen his maritime accomplishments in charting the western, back side of Newfoundland and discovering the Cabot Strait (which John Cabot probably never saw).

In the 1560s other Frenchmen (Protestant Huguenots) attempted settlement a thousand miles to the south, at Port Royal Sound, in what would become South Carolina. Led by Jean Ribaut and sailing under the commission of the admiral of France, the small party staked their colors just to the north of Spanish territory in Florida in the vicinity of Parris Island, and called the place Charlesfort after their king. Ribaut returned to France for supplies, and the small party he left behind fell out with one another and soon perished. France just then was wracked by renewed religious conflict, and Ribaut turned to England, seeking help from the Protestant Elizabeth. He failed. The Spanish meanwhile dispatched an expedition to find the French intruders and root them out (they burned the uninhabited remains of Charlesfort), and when the French Protestants tried again under a veteran of Ribaut's first voyage, slightly to the south at Fort Caroline, the results were no happier. The Spanish then remade Charlesfort into Santa Elena, an actual town and not just a military outpost, where settlers lived until, twenty years later, they were scared off by the marauding English privateer, Sir Francis Drake.

The energies of Elizabethan England that had produced Drake were in the 1580s at last brought to bear on the settlement of the English New World. The person of Sir Walter Raleigh, the great favorite of Elizabeth I, made it happen and partly accounted for its ultimate failure. For this was to be the "Lost Colony" of Roanoke Island, where between 1584 and 1588 a handful of English, including women and children, attempted to found the "Cittie of Ralegh" in the wilderness. Raleigh's efforts on its behalf were mercurial, like his character, and he himself was never permitted by Elizabeth to go there. But he did bring to the enterprise his and his friends' money and all the swagger of the English Renaissance. As a patron he expected, in addition to glory for his name, profit for his investment, and although Roanoke never produced a farthing for anyone, it established expectations that would govern later English attempts to consolidate discovery with settlement. This particular effort fell afoul of court politics and the Spanish Armada, which prevented resupply of the colony at a crucial moment, and when Governor John White finally did return in 1590 he found no English at Roanoke. White lost his daughter and granddaughter (Virginia Dare, the first English child born in America), but England gained experience for when it decided to try again.

When the English did try again, it was in a new reign, and at

Elizabeth I rewarded Walter Raleigh with a knighthood for his efforts in colonizing the New World. (National Archives of Canada)

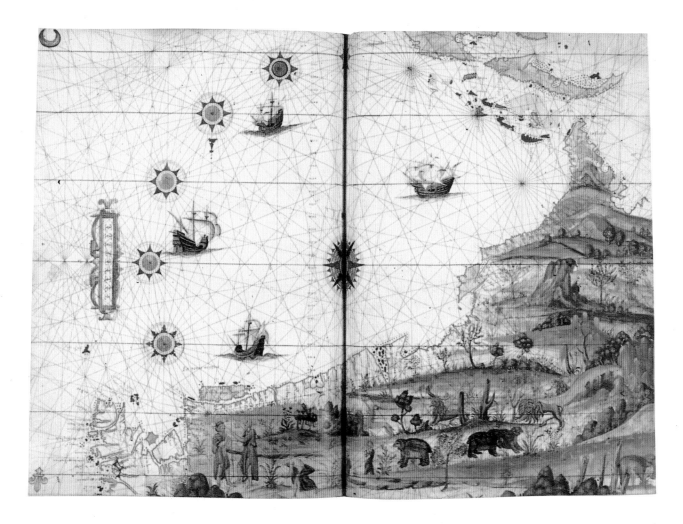

Early maps of the American coast mixed cartographical abstraction with vivid renderings of what the land itself actually held: here, a variety of wildlife. (Koninklijke Bibliotheeck, The Hague)

a much better place, the Chesapeake Bay in present-day Virginia. The result was Jamestown. This time the backing was that of a joint stock company in London, which, personalities aside, was determined to see the colony through to success. Jamestown was founded on a swampy island in the James River in 1607, and its early fitful progress illustrated the still considerable perils of trying to transplant a bit of England to the American wilderness. Bickering abounded, starvation threatened, Indians turned hostile. A surfeit of soldiers and a dearth of farmers crippled early efforts at raising produce that would have been profitable back home, and until the introduction of tobacco, economic prospects seemed dim. But with Virginia the fever for settlement finally caught hold of the English imagination, and Jamestown was the first beneficiary. When that happened, the age of North American discovery came to an end.

Europeans had known about America for decades by the time of the Typus Orbis Terrarum map by Ortelius in 1570, but the seas are still deceptively small, the land immense. (National Maritime Museum, Greenwich)

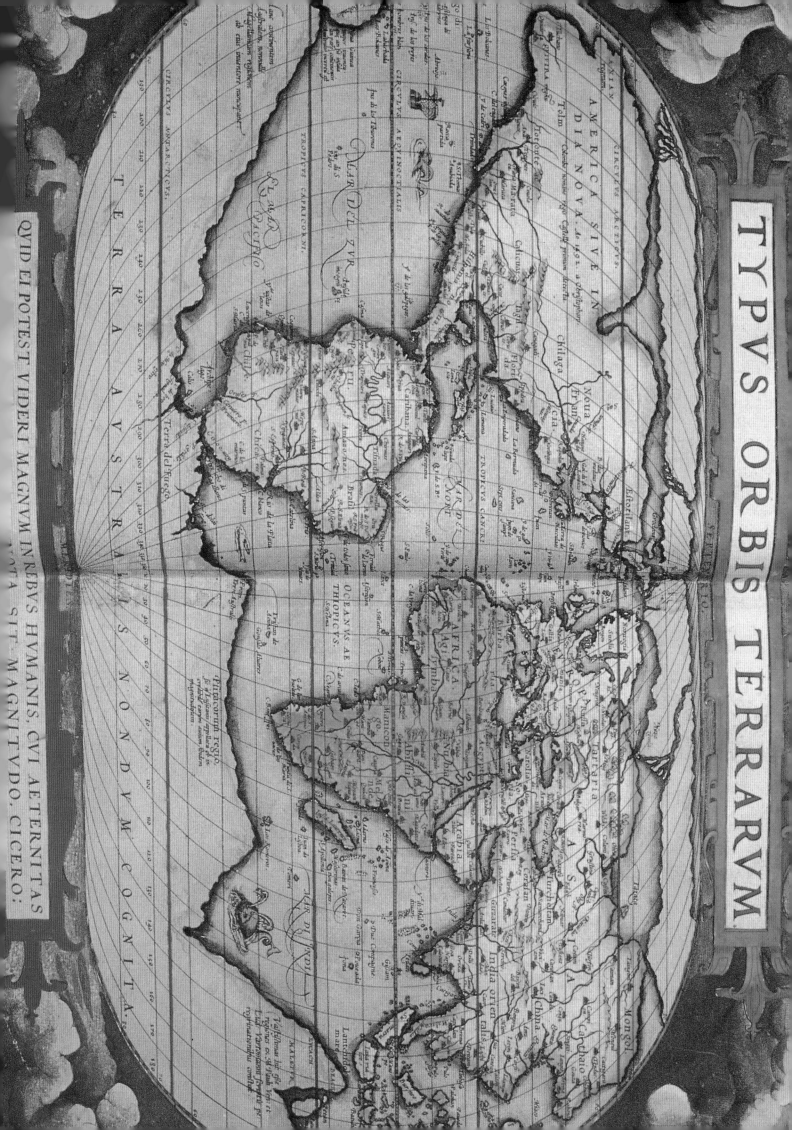

TYPVS ORBIS TERRARVM

QVID EI POTEST VIDERI MAGNVM IN REBVS HVMANIS, CVI AETERNITAS
OMNIS, TOTIVSQVE MVNDI NOTA SIT MAGNITVDO. CICERO:

Sailors and Saints

THE CHRISTIANIZATION OF IRELAND IN THE FOURTH AND fifth centuries by St. Patrick and others opened the door to some of the most myth-shrouded voyages west into the Atlantic. Irish monks, emulating the Desert Fathers of the Near East, sought to remove themselves from the temptations of this world. Monasticism flourished as clerics fled to nearly inaccessible rock islets to live, in tiny cells, what they understood as the pure New Testament life. Their voyages in search of the perfect isolation led to discoveries—perhaps of Iceland, the Faroes, the Azores or the Canaries—and, in the person of St. Brendan the Navigator, to some of the most fabulous tales of Atlantic seafaring. The monks sailed in curraghs, wood-ribbed and skin-covered open boats that accommodated as many as two dozen men, and that were certainly capable of crossing the open ocean. That they ever did in fact reach the North American shore is improbable.

In early Irish literature the voyages were called *imrama*, and the most famous is described in the manuscript *Navigatio Sancti Brandani Abbatis*, which dates from the eleventh century and relates a journey taken by St. Brendan and his fellows supposedly around the year 525. It lasted seven years and was filled with the sort of marvels that gave it a popularity in medieval Britain equal to the legends of King Arthur and the Knights of the Round Table: what they thought was an island turned out to be a giant whale; they came upon a place filled with birds that spoke perfect Latin and that identified themselves as fallen angels of Lucifer. There were magical lights as with Moses' burning bush on Sinai, strange

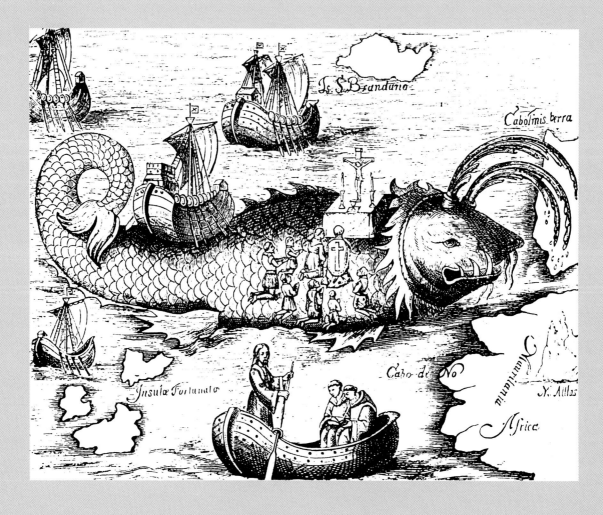

immense objects the color of silver and hard as marble (icebergs) rising out of the sea and a fire-scarred island where there were no trees. On a rock they sighted Judas Iscariot doing penance for the greatest of betrayals and tormented by demons in the company of Pontius Pilate and Caiaphas, the High Priest. In good time, *Terra Repromissionis Sanctorum* or the Promised Land presented itself, a place where there was no night and where a youth greeted the Irish wayfarers singing Psalm 84: "How amiable are thy dwellings, thou Lord of hosts." They were told to enjoy the isle's abundant fruits and to return home with a cargo of precious stones which, as the narrative tells us, they did.

St. Brendan died probably around 580, over ninety years old, safely back in Ireland. The legend of his wanderings deserves a secure place in literature; fanciful and innocent, it would have to be far from the actual experience of monks paddling around the eastern Atlantic in open boats. We do know however that three centuries later the Vikings found Irish bells and croziers—and some monks—in Iceland, and there is good reason to believe that the wandering holy men stumbled onto other Atlantic islands. And they left a record behind them, embellished as it was, that became part of the lore of the Atlantic and that later sailors like Columbus and Cabot (whose motives were rather different from the monks') carried with them on their own voyages of discovery.

The voyages of Irish monks in search of perfect isolation produced some rumors of numerous Atlantic discoveries. The narrative of the most famous of them, St. Brendan the Navigator, told of islands that were whales, and of visions of Judas Iscariot, isolated on a rock and doing penance for the greatest of all betrayals. (Courtesy of the Trustees of the British Library)

Viking Yellers

IT IS DIFFICULT TO KNOW WHETHER TO START OR TO finish with the Vikings. The archaeological evidence leaves no doubt at all that they stepped ashore in North America before the Italians, the Spanish, the Portuguese, the French or the English. Discount the Irish, whose claim to an earlier landing in Newfoundland is unconfirmed, and in the contest for first discovery the prize unquestionably goes to the Norsemen. If you choose to think about discovery as a matter of chronological narrative, then the Vikings are more or less where you would open the story. My own ancestry is Norse, and I can attest to the tenacity of certain characteristics common to the race—appearance, speech, temperament—that can be observed today among the good Scandinavian folk of America's upper Middle West. They are a people who take the phrase "Sons of Norway" with some seriousness, and in the Minneapolis airport on many a weekend you can usually find a group of them, badged and stickered for a "Sons of Norway Tour." They are bound for ten days in the old country, a well-watered place, thickly forested, mountainous, close by the sea, and about as far geographically as you can get from the high, dry plains of North Dakota, which lies landlocked at the very center of the North American continent.

I have made the pilgrimage too, though not the "tour," and have the snapshots of a morning twenty-five years ago spent with an obscure "aunt" in Oslo who someone in the family had figured out was one of us. Like most American "ethnics," Norwegian-Americans today think about the land of their origins with a reverence nurtured by a century's distance, and think not much at all about its place in the chronology of discovery. Yet they bear greater kinship with those shadowy ancestors of a millennium ago who tried to make a go of it in America than they do with my "modern" aunt who never left her comfortable home along the fjord.

Norse mariners left Norway to conquer unknown lands, extending their reach across the North Atlantic. (Terje Rakke/The Image Bank)

For the Norse venturers to North America, the iron sword still represented the pinnacle of weapons technology. Fearsome as such blades were, they fell far short of the percussive firearms that, five hundred years later, made European conquest of the native peoples so irresistible. (Royal Ontario Museum)

If, however, you choose to think about discovery less as chronology and more as a matter of historical memory, then the Vikings are not where you would begin but where you would end up. They are the farthest off; the record of their adventure on our shores is the scantiest; their nature, as a people, is to us the strangest. Plus, they failed to establish a lasting settlement (if, that is, they had ever truly intended to stay). Wherever it was, their colony was, if not romantically "lost," then decisively withdrawn, the memory of it preserved in a rich folk literature but not henceforth used as a practical reference for any further expedition. Not for another five hundred years did Europeans dramatically reach out to possess North America, and these discoverers hailed from Italy and Spain, not the North. Those fifteenth- and sixteenth-century discoverers were as far removed in time from the Norsemen as we are from Columbus. The Norse adventure in America comes down to us through memory at its dimmest, fraught with questionable transmutations and now and then saddled with the outright hoax.

Early in this century, eager and talented charlatans asserted that the Vikings had wandered into America by way of Hudson Bay and then ventured south to explore much of the upper Middle West. Alleged runic inscriptions and Viking relics litter the region. The most notorious—the Kensington Runestone unearthed by farmer Olaf Ohman in Douglas County, Minnesota, in 1898 and made famous by furniture salesman Hjalmar R. Holand—supposedly recorded a Viking visit in the year 1362. Nearby at Alexandria, Minnesota, a larger-than-life replica still draws tourists to Runestone Memorial Park. Near Lake Nipigon, Ontario, a gold prospector produced a sword and an ax—the Beardmore Relics—which, it was claimed, belonged to the wandering authors of the Kensington Runestone (they were acquired by Toronto's Royal Ontario Museum in 1936). Near Newport, Rhode Island, a stone tower built by English settlers about 1675 was said by an authority on the Inca Indians to be an authentic Norse ruin as late as the 1940s.

Dismissed by reputable scholars, the Kensington Runestone and the Beardmore relics are the stuff of the Piltdown Man and the Cardiff Giant. Why such mythology should so profusely surround the Viking adventure to America is hard to say. It represents the kind of willfully mischievous story-telling that can come to surround long-ago events that the popular memory has hopelessly garbled. Such facts as there are about the Vikings come down to us not as deliberate, on-the-spot reporting (such as Thomas Hariot's *A Briefe and True Report of the New Found Land of Virginia*, written on the English expedition to Roanoke Island in 1586), but as works of literature in the form of the two great Vinland Sagas, themselves written down long after the events in question are supposed to have taken place. The Sagas relate those events in the medieval tradition of Germanic heroic poetry: the world of the Nibelungenlied and of Beowulf. They tell of deeds done by people themselves only recently converted to Christianity and in whose minds and souls Thor and Odin still contended with Jesus and Mary. They went into battle without firearms and set out on long voyages without compass. There is nothing modern about them.

But where the actual record is vacant or fanciful, plain plodding archaeology comes to the rescue. To see its results, you must in all likelihood travel a long way. For what the archaeologists have dug up is as remote from us in space as the Vikings are in time. The journey will take you to Canada, to the far northern tip of the island of Newfoundland. When I began the travels that have gone into the writing of this book, L'Anse aux Meadows, Newfoundland, was the only spot I knew for sure I must seek out. For I am Norse, and this is the place, archaeology tells, that proves the Vikings beat Columbus by five hundred years. From my home in Chicago, it is a journey of two thousand miles, and by train and ferry it takes two nights and three days, long enough to make it seem something of an expedition.

It was something of an expedition, I suspect, to Dr. Helge Ingstad, the Norwegian archaeologist who followed the Vikings' trail to this remote end of Newfoundland in 1960. He looked every inch the kindly senior professor off on a dig: a dignified figure in Scandinavian sweater and old-fashioned trousers that laced to the knee, seated on a peat outcropping and talking with George Decker, the local man at whose prompting Ingstad was led to explore some promising-looking bumps and ridges that turned out to hide Viking treasure. What Ingstad and his wife, Anne, found was no hoax.

He had picked up the trail in the Old World, with the migration of Norwegians in the ninth century from their overcrowded homeland to Iceland, which was settled by Ingolf Arnarson in A.D. 870. A tide of Norse and Celtic immigrants soon followed, and by the end of what is called the Age of Settlement, around 930, that rocky volcanic island had a population of thirty thousand and was developing a settled social order and stable political system. A national assembly, the *Althing*, which was both a legislature and judicial body, lay at the heart of a remarkable parliamentary commonwealth that would last for more than three hundred years. Iceland then was part of a rambling, largely unadministered Scandinavian empire that in the tenth and eleventh centuries reached from Russia in the north to Sicily in the south, included much of Ireland, Normandy and the coast of Greenland and extended, briefly, all the way to North America. It was largely a seaborne empire and for the Icelanders in particular sea voyages to faraway places were commonplace. Gradually, both by accidental sightings and deliberate exploration, these transplanted Norwegians were discovering the stepping stones across the northern rim of the Atlantic, which is the easiest way, and was well within the capabilities of tenth-century mariners.

The Norse did not recognize the Atlantic as such, and did not set out across its vast reaches looking for China or some "other"

Sixteenth-century woodcuts from a history by Olaus Magnus reflect the maritime preoccupation of the Norse people. (Library of Congress)

The Norse came to North America not by act of deliberate discovery but by accident in the course of fleeing overpopulated homelands. They first came to Iceland, then it, too, grew crowded.
(John Chang McCurdy/The Image Bank)

world quite confident that they could easily sail home again. No theory propelled them westward, but rather the pressures of population in countries with cold climates and limited arable land. Iceland, not an overly hospitable place, also soon felt the same pressures, which sent some of its heartiest inhabitants westward yet again. They colonized Greenland in 986, as ever in search of pasture for their flocks of cattle and sheep and fertile land for their farms. Icelanders and Norwegians settled on both coasts of that great ice-capped island, and there Ingstad and other archaeologists uncovered abundant evidence of their extensive presence. From the west coast of Greenland, Labrador is only a thousand miles, not a great sail, and it wasn't long before the stout Norse *knarrs* (and not the high-prowed longships of legend) began to call off that remote coast and wander along it. We do not know how many voyages there were. For a time, the northern seas were busy with shipping between Norway and Iceland and Greenland, and the ships may have been to distant Labrador as well. Labrador was heavily forested, and wood was a commodity in short supply in treeless Greenland and in Iceland. The place was supposedly bumped into by Bjarni Herjolfsson, who while en route in 986 from Iceland to Greenland was blown off course and came across, somewhere to the west, a relatively flat and wooded country where no glaciers covered the land (as they did in Greenland). Singularly incurious, he did not go ashore but returned to Greenland and eventually to Norway, where he died prosperous—and in bed. But word got about, and soon others followed where he had gone.

The most famous was Leif Eriksson, also known as Leif the Lucky, whose father, Eric the Red, had led more than a dozen shiploads of Icelandic settlers to Greenland's west coast in the 980s. Leif struck out on his own (in the boat he had bought from Bjarni) in the summer of 1001 and, as the Sagas tell it, was soon cast up on a far shore—first probably on Baffin Island, then on Labrador (which he called Markland, land of forests) and then on another land several days' sail to the south. They probably next set foot on Belle Isle, in the narrow strait between Labrador and Newfoundland that is ice-free only in the summer months, and then moved on to a harbor on what appeared to them to be the mainland but was actually the northern edge of Newfoundland, near Cape Bauld. They beached their *knarr* and inspected what soon showed itself to be a green and pleasant land: sweet streams filled with fat salmon, lush meadows, ample trees—and no ice. The description more or less fits L'Anse aux Meadows on Épaves Bay. Leif's crew built shelter, cut timber, gathered grapes and spent the winter. The "grapes" may have been blueberries, which still grow in the area, or some other variety of vine that one thousand years ago, when the climate was warmer, might have grown at this latitude. In the spring they all returned to Greenland with their cargo of lumber and raisins or wine—the evidence that they had found a new land. They called it Vinland.

They left no one behind, but we can surmise from their profitable cargo that they planned to return. There are Saga records of three more expeditions to Leif's site in Vinland. Leif's brother Thorvald went there in 1004–05 and became the first European of record (and it is a Saga record only) to be killed by the American aborigines. In 1008 Thorfinn Karlsefni, an Icelander and close friend of the children of Eric the Red, set out with a colonizing expedition of three ships and 250 men and women, plus livestock; they spent one winter at Leif's settlement and another somewhere along Newfoundland's west coast, before throwing in the towel to the increasingly hostile natives and heading home for Greenland. Finally, in 1013 and 1014 two brothers from Norway named Helgi and Finnbogi, along with Thorvard of Gordar, a Greenland farmer, and his wife, Freydis, the illegitimate daughter of Eric the Red, sailed for Vinland, agreeing to share the profits among them. But Freydis and Thorvard murdered the two brothers and their crews, gave the place up and, back in Greenland, got off with only a curse from Leif.

All of these voyagers seem to have spent time at the same spot in a land they all called Vinland. That spot may have been L'Anse aux Meadows. Or L'Anse aux Meadows may only have been a way station en route to some other settlement that has yet to be unearthed. But the general clues from the Sagas conform to the

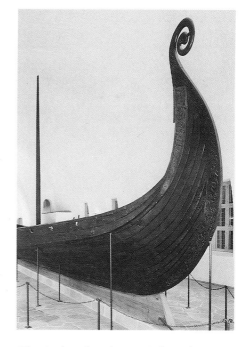

The Oseberg longship in Oslo is the most famous of Viking vessels, though a replica, which actually sailed the Atlantic in 1893, can be seen in Chicago. (University Museum of Antiquities, Oslo)

location and character of L'Anse aux Meadows, and it remains the only North American site where archaeology definitely confirms a non-native human presence—likely Norse—at the right period of time. Ingstad had spent years excavating Norse sites in Greenland, and it was there that he theorized that the location of Vinland was much farther north than earlier conjectures, which had put it in the vicinity of present-day New England. L'Anse aux Meadows proved him correct.

Parks Canada now manages the site at L'Anse aux Meadows and provides just about everything you would want to know about Leif, Thorvald, Thorfinn and the rest. It is located twelve miles from the town of St. Anthony on Newfoundland's Great Northern Peninsula, at the end of the highway from Deer Lake. The development and tourism people have dubbed this road, hopefully, The Viking Trail and posted it with signs bearing a likeness of dragon-prowed longships and horned helmets of the sort worn in Hollywood Viking epics. Displays on the ferries from Nova Scotia and in airports and bus stations offer the standard promotional brochures and gimmicks designed to lure tourists several hundred miles north to this most remote part of Newfoundland. Summertime abounds in those "special events" that local chambers of commerce like to think make for "extra-special memories": Salmon Festival at Plum Point, shrimp shuckings at Port au Choix, bake-apple tasting in Forteau, Cod Festival in St. Anthony, Lobster Days at Cow Head. "Don't be surprised to see a Viking dressed in costume. He'll share the Viking culture and folklore with you. Join in a 'yell-in' . . . maybe you'll become a member of the Royal order of Viking Yellers!"

Happily, the place is as spectacular as the promotion of it is banal, and I do not hesitate for a moment to say that it amply repays the considerable time and effort it takes to make a visit.

Inside the L'Anse aux Meadows Visitor Center you will be greeted by earnest, professional and bilingual Canadian park service employees, likely as not from Sarnia or Saskatoon. They will try to answer your questions in English or French, will show you a movie and take you on a tour. There is a handsome exhibit re-creating the act of saga-telling in a spooky replica of a Norse lodge, and much material on the aborigines who were here first. The film informs and entertains with a "you-are-there" newsreel-type rendition of Ingstad's long efforts to corroborate the tales told in the Sagas. One senses in these thirty-year-old clips something of the awe that this grandfatherly modern-day Norseman must have felt for his long-lost Viking adventurers. At a well-stocked book stall you can buy a paperback copy of the Sagas and compare notes as you walk about.

What Ingstad found, when he found it, must have been pretty thrilling, but as is the way with real archaeological research (unlike

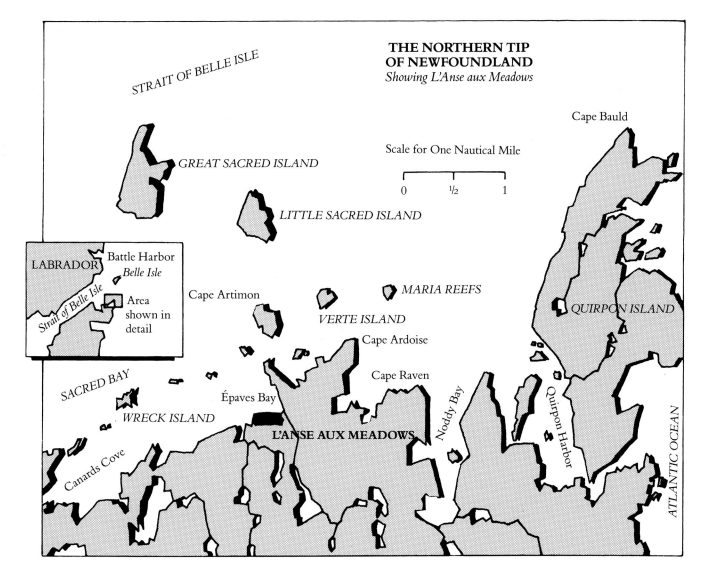

**THE NORTHERN TIP
OF NEWFOUNDLAND**
Showing L'Anse aux Meadows

the storybook variety where old maps lead to buried treasure and where *X* always marks the spot), it yielded in the end not a great deal to look at. Ingstad was joined in the excavation by archaeologists from Norway, Iceland and the United States (the Canadian park service later followed up with its own digs), and together they found, just where local George Decker said they would, the lower courses of the walls of eight Norse structures dating from the eleventh century and of a sort then common in Greenland and Iceland. The buildings would have had sod roofs supported by wooden rafters; inside, in the middle of a clay floor, there would have been long firepits for heating, cooking and illumination. Doors would have hung on leather hinges. Two of the largest buildings closely resembled those Ingstad had grown familiar with from his Greenland digs. About fifty-five by seventy feet, they probably served as a kind of communal dwelling for the settlement: the place to eat, sleep and talk away the long northern winter. Half a dozen smaller rooms appear to have been clustered around the central hall, some with their own center hearths. Seventy-five to ninety people, Ingstad judged, were accommodated here at any one time.

For seasoned sailors, the tenth-century Norsemen who settled briefly at the site of L'Anse aux Meadows in northern New-foundland picked a remarkably inhospitable harbor at Épaves Bay. (Dudley Witney)

The L'Anse aux Meadows excavation archaeologically confirmed what the literature of the sagas said: that the Norse found and for a time settled an immense land to the west of Greenland, which they called Vinland. (R. Ferguson, Canadian Parks Service)

Whereas the existence of easily degradable sod buildings from one thousand years ago was almost, if not quite, a matter of inference, the discovery of other small artifacts of whose provenance there could be little doubt seemed to clinch the case for L'Anse aux Meadows. There were the makings of what might have been a woman's sewing kit (the Sagas said Norse women had gone to Vinland) with soapstone spindle whorls, knitting needles, a pair of small scissors, and a whetstone to sharpen them. But Ingstad's diggers knew they had hit home when they came upon an iron boat nail of the sort commonly used in Norse cultures to join together the clinker-built hulls of their *knarrs* and longships. Some of the nails may have come from ships made in older Norse settlements, but many apparently were forged on the spot. Large amounts of slag, produced in the melting and smelting of bog iron, suggest the existence of a furnace and a forge, probably beside Black Duck Brook near where it flows into Épaves Bay. It was surely a primitive thing like those known in Greenland and Iceland—no more than a clay-lined pit with a lid of large stones where iron ore from local peat bogs was roasted over charcoal to temperatures around 2200⁰ Fahrenheit and then hammered out over a smithy to remove impurities. It was not an efficient process, for the remaining slag retains four-fifths of its original iron. The later Canadian park service excavations also revealed a large amount of debris associated with the preparation of planking with metal tools: no doubt the lumber referred to in the Sagas and that was so sought after back in Greenland. It also suggests that the settlers almost certainly engaged in boat building and repair; this is a place where boats can easily be hauled ashore and careened.

It all adds up to the remains of a small village. If this is the place referred to in the Sagas, then some of the buildings were probably Leif's, and some were added by his friends and family who came out over the next few years. We do not know if any of these people could read or write, only that the site has yielded no runic inscriptions: no literal messages remain to speak from that time to this. Charred roof timbers indicate the whole place was burned down after the Norse gave up and went back to Greenland, and carbon 14 dating of the remains yields an age a century on either side of the year 1000, which is when the Sagas say Leif and the others were there.

Parks Canada knows its job, and part of its job is to attract an audience. L'Anse aux Meadows is probably never going to be like Yellowstone or Banff; I visited in August and was not much disturbed by the crowds. But if my rude survey is any measure, what intrigues most visitors once they're there is less the actual archaeological ruins, which are just small grassy hillocks suggesting squares and rectangles, than the painstaking reconstructions that sit nearby. *Come In; Please Do Touch; Imagine Yourself a Viking in America a Thousand Years Ago; Have a Friend Take Your Photograph Beneath a Viking Lintel.* There are as yet no costumed "interpreters" with scripts at the ready, just a skiff and three replica buildings made of thick peat bricks and roofed with sod, half sunk into the ground. From a distance they look like hillocks themselves.

The reconstructors surely have it right. This is what Norse shelters looked like in Norway, Iceland and Greenland at the time and for that matter what they still looked like on the plains of North and South Dakota less than a hundred years ago. Warm and quite weatherproof, they served the needs for shelter of a simple people. Today they serve, as do all such honestly done historical re-creations, to assist the modern imagination back into a remote culture. In this, the presence of even a few other visitors is vexing, but it is hard to have it both ways. Re-creations are fun in a way that ruins aren't, particularly ruins that are hardly there at all.

That history didn't really happen inside these huts constructed in the 1970s is easily forgotten, I suspect, by many visitors, so authentic is their detail and their feel, right down to the ashes in the firepits and the shaggy fur sleeping bags that obviously could use a good wash. The air smells pleasantly of smoky peat (an odor much like that of scotch whisky) and not of a crowd of unbathed Vikings (but there is little ventilation so you can easily get the idea). There is no graffiti of the "Leif Loves Freyda" variety, though there is certainly ample place and opportunity for it. Because the houses are empty of inhabitants, the mind's eye naturally fills them with spirits of its own, and the ghosts of Leif Eriksson or Thorfinn Karlsefni are easily summoned. At this particular site, the keepers

The reconstructions of Norse buildings at L'Anse aux Meadows attract most visitors. Mere hillocks mark the actual ruins. (B. Schonback, Canadian Parks Service)

can also safely count on the romantic allure that attaches to places whose origins and fate are only partly known—places that exist, tantalizingly, both in the historical record and outside of it in the realm of the imagination and perhaps even make-believe. We don't really know who these people were who came to this particular place, nor do we know for sure what became of them. Probabilities connect an archaeological record with a literary one, but they leave room enough for mystery.

All this quiet, low-key authenticity sits against a natural backdrop unchanged since the Vikings' visit: shallow, shoal-filled, north-facing Épaves Bay, not so much a bay as a broad curving beach, looks out past Wreck Island, Great and Little Sacred islands, across the Strait of Belle Isle to Belle Isle and Battle Harbor in Labrador. To the landward, peat bogs, moors and scrub timber rise past countless lakelets and ponds back into the interior of the Great Northern Peninsula. Modern times intrude with a few power poles and sodium lamps of adjacent settlement, and with the park Visitor Center itself.

Above it fly three flags, all unquestionably modern: Canada's red and white maple leaf, Newfoundland's abstract ensign of white (symbolizing snow and ice), blue (the sea and the commonwealth heritage), red (human effort), and gold (confidence), and the United Nations' white globe on a blue field. This last is not an idle gesture of goodwill to foreign visitors. It is a token of the designation of L'Anse aux Meadows to the World Heritage list of the United Nations Educational, Scientific and Cultural Organization

The Norse technique of building with sod, seen here in the L'Anse aux Meadows reconstruction, came with them from Norway, and was not forgotten by their nineteenth-century descendants on the great plains of Canada and the United States. (J. Steeves, Canadian Parks Service)

The sod-and-pole reconstruction of the Norse settlement at L'Anse aux Meadows makes it easy to imagine the nature of tenth-century life on the Newfoundland shore. (B. MacDonald, Canadian Parks Service)

(UNESCO). "An outstanding cultural site forming part of the heritage of mankind," says the plaque. "L'Anse aux Meadows is the first authenticated Norse site in North America. Its sod buildings are thus the earliest known European structures on this continent: its smithy the site of the first known iron working in the New World; the site itself the scene of the first contacts between native Americans and Europeans. It is therefore one of the world's major archaeological sites."

L'Anse aux Meadows is everything that the Kensington Runestone was not. It exudes the very air of scientific truth and is brought to the world with all the measured correctness that the civil servants in Ottawa and their field workers can muster. The site evokes a mood quite different from all the hoopla about Viking yell-ins out along the Viking Trail. One wonders which treatment old Leif would have preferred.

If this really is the "first" site and so the most distant from us in time, then its remoteness in Newfoundland seems especially fitting, for Newfoundland is also just about as close to the Old World as the New World gets. This fact reminds us that transoceanic distances truly were daunting to the best of mariners five hundred and a thousand years ago and that the ocean was seen at once as an obstacle to some discovery and as the path that would lead back home. If ever that path were lost, that connection broken for long, then European settlement in America could not survive. This was what happened with the Norse colonies in Greenland during the fourteenth century, and more precipitately to the English colony at Roanoke Island at the end of the sixteenth century. The physical remoteness of the place, even as experienced today, underscores the precariousness surrounding the events that occurred here and is a reminder of how easily it all could and sometimes did come undone. But perspective demands we remember that the precariousness that to us seems quite awesome, used as we are to the

The Norse settlement at L'Anse aux Meadows lacked the shelter of Norway's fjords, but the climate was mild and the salmon were fat. (B. Wallace, Canadian Parks Service)

disciplined regularities and securities of modern life, was to the Norsemen and most other human beings living a millennium ago the rule of daily existence, and that the hardships, fears and loneliness they endured on the far coast of North America were different only in degree (and not necessarily great degree) and not in kind from what life offered in older, more familiar worlds.

On one subject out of the many that are part of the total story, modern interpretation of these premodern events comes closer to capturing the essence of the discovery experience than all others. This is the subject of the aborigines whom the Europeans confronted on these shores and whose way of life was ultimately forfeit in the encounter. The Norse experience in this regard anticipated the sad pattern repeated again and again in the later chapters of the discovery story. The intruders' and the natives' first discovery of each other was marked by a benign moment of curiosity and by almost playful trading of trinkets for skins and food. But it was soon followed by suspicion, hostility, ambush, reprisal, subjugation, disease and death.

Historians today bend over backward to display sensitivity to the native victims of discovery. This inclination dates to the 1960s' enthusiasm for cultures that were nonwhite and non-Western, and it has resulted in a rummaging of the past for neglected or oppressed peoples in need of rehabilitation. To it we owe the phrase "Native American," which was offered up to replace Columbus's "Indian" as a satisfying non-European, nonethnocentric description of the dark-skinned people who were here before the white men came. At L'Anse aux Meadows, everything is very au courant in this respect also. The archaeologists tell us that native peoples had used the site for some six thousand years, and in the surrounding area scholars have excavated Stone Age implements and encampments indicating the presence of at least six different groups. Two hundred years before the Norse, Dorset Eskimos apparently lived on Épaves Bay, and at Port au Choix on Cape Riche farther to the south is another Parks Canada site devoted exclusively to them. There is also evidence that a group of Indians had migrated from the Labrador side of the Strait of Belle Isle and were at or around L'Anse aux Meadows at the time of the Norse. The Beothuk Indians had lived in Newfoundland since the third or fourth century A.D., moving between seashore and forest with seasonal changes in the food supply. In winter they lived communally in substantial log houses, in summer in flimsy skin-and-bark tents. We know they crafted birchbark into toys and canoes and sewed moccasins and clothing from caribou skins. It may have been Beothuks who came upon the Vikings and so set in motion the familiar pattern of native–white relations.

What we know of that confrontation comes from the Sagas and must be weighed against the character of that source. A mixture of fact and fancy, history and invention, Christian present and pagan past, they are documents concerned basically with people whom the saga writers saw as involved in a great adventure. Their subject is roughly historical, their context literary. Both Sagas that treat the Vinland voyages—*Groenlendinga Saga*, written about 1190, and *Eirik's Saga*, a deliberate refinement of the earlier work and written about seventy years later—tell us something about the natives. In their coarse and vigorous language we catch the stark simplicity of an encounter between human beings who no doubt barely recognized one another as such.

The natives are first reported during the second summer of Thorfinn Karlsefni's colony, and the Norse term used to describe these odd people, "Skraelings" or "Skrellings," captures the nature of the relationship. It is a term of contempt meaning wretches, weaklings, barbarians, even pygmies. "A great number of them came out of the wood one day. The cattle were grazing nearby and the bull began to bellow and roar with great vehemence. This terrified the Skraelings and they fled, carrying their packs which contained furs and sables and pelts of all kinds. They made for Karlsefni's houses and tried to get inside, but Karlsefni had the doors barred against them. Neither side could understand the other's language."

Small-time trading ensued, the milk from the Norse cattle being the natives' chief object of desire after weapons, which the Norsemen would not give. The natives returned "early next winter in

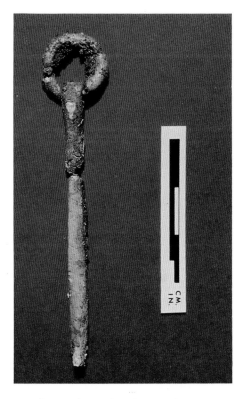

Not history but archaeology confirms the presence of Norsemen on the northern Newfoundland shore: here, a bronze pin. (G. Vandervloogt, Canadian Parks Service)

Remains of a Norse sword, ax head and rattle—the artifacts of a vanished northern culture whose accidental and glancing encounter with North America left little else behind. (Royal Ontario Museum)

much greater numbers this time," there was more milk-for-furs bartering, and one native was killed "by one of Karlsefni's men for trying to steal some weapons." At their third, and now openly hostile, visit, the Norsemen lured them to battle on an open spot between a lake and a wood: "The fighting began and many of the Skraelings were killed. There was one tall and handsome man among the Skraelings and Karlsefni reckoned that he must be their leader. One of the Skraelings had picked up an axe, and after examining it for a moment he swung it at a man beside him, who fell dead at once. The tall man then took hold of the axe, looked at it for a moment and then threw it as far as he could out into the water. Then the Skraelings fled into the forest as fast as they could, and that was the end of the encounter."

In *Eirik's Saga*, we read that the Skraelings first approached in skin-covered boats to trade peacefully, and that "they were small and evil-looking, and their hair was coarse; they had large eyes and broad cheekbones." A nine-inch span of red Viking cloth brought the Norse, in exchange, one grey pelt. Several weeks later, however, Karlsefni's men spied a huge number of boats "pouring like a torrent," their occupants bent on battle. The Norsemen obliged and, as we learn in a famous bit of earthy Saga detail, were saved from the savages only by the bared breasts of a pregnant woman: "When they clashed there was a fierce battle and a hail of missiles came flying over, for the Skraelings were using catapults. Karlsefni and Snorri saw them hoist a large sphere [an inflated moose bladder] on a pole; it was dark blue in color. It came flying in over the heads of Karlsefni's men and made an ugly din when it struck the ground. This terrified Karlsefni and his men so much that their only thought was to flee, and they retreated farther up the river. They did not halt until they reached some cliffs, where they prepared to make a resolute stand. Freydis came out and saw the retreat. She shouted, 'Why do you flee from such pitiful wretches, brave men like you? You should be able to slaughter them like cattle. If I had weapons, I am sure I could fight better than any of you.' The men paid no attention to what she was saying. Freydis tried to join them but she could not keep up with them because she was pregnant. She was following them into the woods when the Skraelings closed in on her. In front of her lay a dead man, Thorbrand Snorrason, with a flintstone buried in his head, and his sword beside him. She snatched up the sword and prepared to defend herself. When the Skraelings came rushing towards her she pulled one of her breasts out of her bodice and slapped it with her sword. The Skraelings were terrified at the sight of this and fled back to their boats and hastened away." But despite Freydis's heroic stand, enough was enough: "Karlsefni and his men had realized by now that although the land was excellent [that winter there had been no snow at all, and livestock had been able to fend for themselves], they could never live there in safety or freedom from fear, because of the native inhabitants. So they made ready to leave the place and return home." En route, they came upon five Skraelings camped on a beach and, with feelings still running high, promptly dispatched them.

Their decision to leave was no doubt a sound one. They were not after all part of some larger colonizing effort and could not count on an ever-increasing flow of new recruits to reinforce and secure their foothold. And their weapons were not yet superior enough to make up for the great numerical imbalance of forces. Nor were they prepared by their past experience to deal with hostile natives. In Greenland they had confronted none at all, for there the Eskimo natives had moved far north as the climate had warmed during the tenth and eleventh centuries. When the temperature fell again in the fifteenth century, the Eskimos reappeared and likely hastened the end of Norse Greenland, which by the time southern Europeans rediscovered North America in the 1490s had returned to a land of fog and ice. Vinland too, with its victorious Skraelings, was left in peace for another five centuries.

The battles between the Skraelings and the Vikings were just the first of thousands of such fights between natives and Europeans that began again in 1492 and did not end until 1890 at Wounded Knee, South Dakota, which witnessed the last forlorn armed resistance of North American Indians to the white man's world. We know essentially nothing of the mental and psychological state of the Skraelings as they first did battle with the Norse intruders, and it would be perilous perhaps to infer too freely backward from what we do know about the state of mind of their spiritual descendants, the Sioux and the Apache, who centuries later fought their last doomed battles on the Great Plains in the deserts of the Southwest with troopers of the United States Cavalry.

Doomed too were those nameless "wretches" who drove off the Vikings from ancient Newfoundland, though they surely didn't know it. Whether the Skraelings preserved any folk memory of the Norsemen with their milk and red cloth and iron knives and fierce women is equally unknowable. Their only hope lay in isolation, which even these tentative Viking voyages proved might not last. When the isolation of North America was violated, permanently, beginning with Columbus, it was by Europeans in possession of more powerful arms and possessed by more powerful ideas than those of the scruffy little heroes who had pushed the mighty Vikings back into the sea.

A confrontation between a Norseman and a Skraeling, as depicted in a woodcut by sixteenth-century historian Olaus Magnus. (British Museum)

The Vikings were a feared and hardy race, and chief among them was Eric the Red, as shown in this engraving. (Arnamagnaen Collection, Aarhus, Denmark)

From Columbus to Cabot

BETWEEN THORFINN KARLSEFNI AND JACQUES Cartier, there were several European voyages to Canada that are a matter of some kind of record. All were glancing blows, quick thrusts quickly withdrawn: John Rut's from England in 1527; Giovanni da Verrazzano's from France between 1524 and 1528; João Fagundes' in the early 1520s and Gaspar Corte Real's in 1500 and 1501, both of whom were from Portugal; João Fernandes from the Azores, also around 1500. The paucity of the record, however, leaves a certain dimness about the history of these early episodes, as it does about the first of them, John Cabot's voyage from England in 1497. Dimness is the point. John Cabot, like Columbus five years before and farther to the south, was looking for a short route to the Indies. He neither tried to stay in North America (as Cartier unsuccessfully did) nor did he lead the way for others who soon would make the New World their permanent home (as the Roanoke voyagers did for their Jamestown successors). Cabot made just one and a half voyages, with only one ship—the "half" because sometime during the second voyage, the Atlantic claimed him. If the story of the discovery of the North American continent can be said to have begun with him (Columbus never saw North America), then the lesson to take from it lies with the fact that it was a very tentative beginning, from which more fulfilling ventures neither quickly nor necessarily followed.

Perhaps because he offers the opportunity of proving the "first landing," Cabot has been the subject of relatively greater scholarly effort than the others in these early dark years. The bibliography and notes at the end of the chapter on Cabot's voyages in Samuel Eliot Morison's *The European Discovery of America: The Northern Voyages* (Oxford, 1971) run for nineteen pages. All the known sources took James A. Williamson fifty years to compile and interpret, and they fill a book to themselves: *The Cabot*

Diego Ribero's world map of 1529 revealed the beginnings of the Spanish Empire in the New World, which by the end of the sixteenth century made Spain the richest nation in Europe. (Biblioteca Apostolica Vaticana)

41

Woodcuts eventually embellished Columbus's reports of his voyages for his patrons and audience back home. They suggest how a new Spain might be built on the shores of a new world. (New York Public Library, Astor, Lenox and Tilden Foundations)

Voyages and Bristol Discovery under Henry VII (The Hakluyt Society, 1962). Morison, who probably read every word of it, characterized the research as both "intense and assiduous," but with results that were "disappointingly meager." There is no portrait or even description to tell us what John Cabot looked like. There is no log of his voyages, no account of them directly from his sailing companions, no sample of his handwriting, no signature. From a jumble of thirdhand hearsay, rumor and guess, Morison assembled what can be regarded as the definitive account.

Cabot was probably a Genoese, like Columbus, and the name may have been spelled accordingly: Cabotto, Chiabotto, Savoto. It means "one who engages in coastal sailing." He was born in the early 1450s, and before he was ten moved with his father to Venice, where he married and lived through the 1480s. There is archival evidence that a John Cabot Montecalunya, a Venetian, lived in Valencia, Spain, between 1490 and 1493 and that he approached King Ferdinand about building a jetty, something that apparently never came to pass. This man, if it was Cabot, might well have witnessed the triumphal arrival of Columbus in Barcelona in April 1493, home from his first rousing voyage to the Caribbean. Or was it China?

Seville and Lisbon also seem to have heard the petitions of a Cabot who said he could find a shorter route yet than Columbus's. If at this point Cabot was a would-be Columbus, frustrated that the other Genoese (whom he could conceivably have known in childhood) had found the first new and shorter route to the spice-rich East by sailing west, but still stubbornly determined to blaze his own path between the continents, then it is plausible that the next best place for him to look for support was England. England in the 1490s was at peace with herself, the Wars of the

Roses having ended on Bosworth Field in 1485 and her crown now securely in possession of the new Tudor dynasty. England also sat at the far end of the spice route from the Indies, which then still stretched eastward through the Mediterranean Sea and the Indian Ocean with various overland connections. Demand was high, the supply inadequate. Columbus had figured this out too, and in his own long and discouraging campaign to find the patronage that would enable him to give it a try, he had paid a call on the new English king, Henry VII. Henry, frugal Welshman that he was, turned Columbus down but lived to regret it when, flying Spanish not English colors, the Genoese made his great discovery in 1492.

It is one of the more intriguing "ifs" of discovery history: if Columbus had sailed for Henry instead of for Ferdinand and Isabella, and if he had taken the same southerly course across the "Ocean Sea" toward what he fairly believed was the Asian mainland, and if he had in fact found the same Caribbean islands and then the coast of Central and South America, then would "Latin" America have become "Anglo" America? And what would have become of the actual "Anglo" (and for a time also partly "Franco") America that lies to the north of Florida? Would Spain, jealous of England's new riches and determined not to lose out in the New World entirely, have pushed to compensate by colonizing the lands that later became the United States and Canada? Would southern America then have become the place where liberal democratic institutions took root and flourished, to become in our own age bastions against modern totalitarianism? Would northern America then have become a fractured land of contending nationalisms and authoritarian habits? Would southern America have become the Jeffersonian home to religious and political toleration but racial intolerance? Would northern America have become solidly Roman Catholic, utterly unbending in matters of the faith but harboring a remarkably large store of tolerance on matters of race? Or would Spain have done nothing at all (just as England and France did nothing much for over a hundred years after Columbus) and have left the whole field to the northerners? Might there then have developed not two Americas, but one?

It is amusing to speculate, but it is probable that even if the tables had been totally reversed we still would have ended up with two Americas, which despite much loose talk about the New World (singular), in this respect precisely mirrored the Old World—riven north-to-south along lines of religion, culture and political temperament. The United States and Canada are lands sprung from the anti-authoritarian tradition of *Magna Carta*, of John Locke's *Second Treatise on Civil Government*, Jean Jacques Rousseau's *Social Contract* and Adam Smith's *Wealth of Nations*; they are animated by the contentious temper of Luther, Calvin, Cromwell and Burke. Neither Mexico nor any nation to the south shares that heritage. Judged, then, by the two distinct Americas that in fact subsequently came about, both Columbus and Cabot

This woodcut shows a ship similar to the Santa Maria. *No contemporary drawing of Columbus's ship has survived. (British Museum)*

are due equal billing in the sweepstakes of discovery: they discovered new worlds destined to become as different from one another as they were from the old. (And neither new world ended up speaking Italian.)

So John Cabot, who had moved with his wife and family to Bristol in the west of England in 1495, petitioned Henry VII and, providing he could pay his own way, got to write his own ticket. Modeled on Portuguese and Spanish precedents, Cabot's letters-patent from Henry are dated March 5, 1496, and their language captures the adventurous, wildly inclusive aggrandizing spirit of the age of discovery just then about to begin. Granted to "our well beloved John Gabote, citizen of Venice, [and] to Lewes, Sebastian, and Santius, sonnes of the said John . . . full and free authoritie, leave, and power, to sayle to all partes, countreys, and seas, of the East, of the West, and of the North, under our banners and ensignes, with five ships . . . and as many mariners or men as they will have with them in the saide ship, upon their own proper costes and charges, to seeke out, and discover, and finde, whatsoever iles, countreyes, regions or provinces of the heathen and infidelles, whatsoever they bee, and in what part of the world soever they be, whiche before this time have beene unknowen to all Christians."

His base in Bristol, since 1400 the second-largest seaport in England, was well and deliberately chosen. Bristol lies midway between Iberia and Iceland and in the late Middle Ages built its substantial prosperity on a north–south trade involving English wool and woolens from the Cotswolds, dried cod from Iceland, and olive oil and wine (chiefly sherry, thus "Bristol Cream") from Spain and Portugal. Bristol fishermen had sailed to the west for an unknown number of years in search of a new base and, who knows, may have stumbled onto the far coast of North America in advance of Cabot. But nothing is recorded, and if previous discovery were strongly rumored it seems unlikely that Henry would have granted to Cabot the broad license he did. But that very imperial-sounding charge was accompanied by nothing else—"Upon their own proper costes and charges"—which left Cabot to find financial backing on his own.

The expedition was correspondingly small. Henry said five ships could go under his banners. One went: the *Mathew*, a *navicula* (little ship) of just fifty tons burden, probably square-rigged on the fore- and mainmasts, with a lateen sail on the mizzen. There were eighteen crew, all English except for a barber from Castiglione near Genoa and one Burgundian. They set off in the spring of 1497, floating down the River Avon from Bristol, to the Severn Estuary, into the Bristol Channel between the mountains of Henry's Wales and the moors of Somerset and Devon, around Dursey Head on the southwest coast of Ireland, and out into the open Atlantic. Keeping the North Star on his starboard side, Cabot sailed along latitude 51 degrees 33 minutes north, and thirty-two or thirty-three days later he sighted Newfoundland; he took the same amount

The compass was every captain's most valuable navigational instrument and, for two centuries before Columbus, navigators had used it out of sight of land to avoid getting lost. Here, an Italian example, circa 1580. (National Maritime Museum, Greenwich)

Henry Tudor, the Welshman who ended the Wars of the Roses and brought peace to a united England, sponsored John Cabot's voyage in frugal fashion. Cabot took only one ship, the Mathew. (National Maritime Museum, Greenwich)

of time as Columbus, who had sailed along 28 degrees north from the Canary Islands, but on the northern route he went only half the distance. At latitudes north of 40 degrees, thirty-three days is a very respectable crossing, and Cabot's record stood for years to come.

From the inscription on Sebastian Cabot's map of 1544 (Cabot's son who may or may not have been along on the trip), we know landfall was made at 5:00 A.M. on June 24, and Morison has concluded that the land they saw was Newfoundland's Great Northern Peninsula at Cape Dégrat, which rises to the height of 502 feet above the sea and would have been visible from the *Mathew*'s masthead at a distance of twelve to fifteen miles. Cabot sighted a large island to the north, which he called John the Baptist, after the day (the French later called it Belle Isle, which is still its name), and so he must have seen the entrance to the northern

Stout Wooden Ships

THE VOYAGES OF DISCOVERY WERE UNDERTAKEN WITH the tools at hand, and the ships available by the end of the fifteenth century turned out to be well-suited for the job. Mostly they were stubby merchantmen, designed and built for the rough and tumble of trade in Europe's home waters. These were ships that routinely tackled the English Channel, the Bay of Biscay and the Baltic, waters as challenging as the North Atlantic, only smaller.

Distance rather than the fearsomeness of the seas raised the risks of discovery, for given the faulty charts of the times many a small vessel ventured out only to find more water where there should have been land. Broad beams relative to their lengths made for vessels of great strength and seaworthiness, although they were of little comfort. John Cabot's *Mathew*, Jacques Cartier's *La Grande Hermine*, Christopher Newport's *Susan Constant*, were the products of the combined evolution of northern and southern traditions of shipbuilding. From Scandinavia came the square-rigged *knarr* or longship, a swift and sturdy design with small cargo capacity, but which by the fourteenth century was developing stouter lines, heavier construction and fore and stern "castles." From the Mediterranean came the lateen-rigged ship made for short-haul commerce, a stout ship with triangular sails that enabled it to sail closer to the wind (and therefore the shorter distance) than the square-rigged *knarrs*. The two styles came together in the carrack, which by the time of Columbus and Cabot carried two well-stayed masts, with square sails on the mainmast, and a lateen sail on the mizzen.

This model of a fourteenth-century ship from the Mediterranean illustrates the lateen rig, which found its way onto later designs, usually on the aft or mizzen mast. By the age of discovery, steer-boards (on the right side of a ship, thus the term "starboard") had given way to proper, center-mounted rudders. (Science Museum, London)

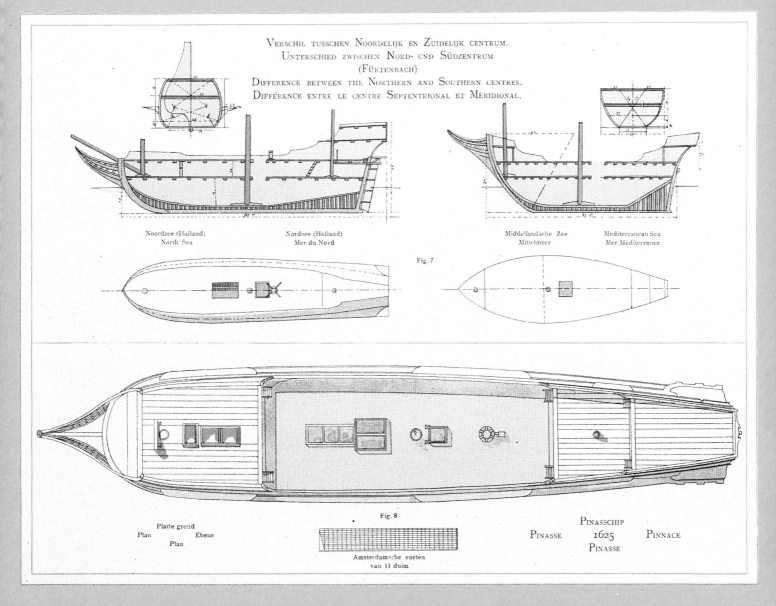

VERSCHIL TUSSCHEN NOORDELIJK EN ZUIDELIJK CENTRUM.
UNTERSCHIED ZWISCHEN NORD- UND SÜDZENTRUM
(FÜRTENBACH)
DIFFERENCE BETWEEN THE NORTHERN AND SOUTHERN CENTRES.
DIFFÉRENCE ENTRE LE CENTRE SEPTENTRIONAL ET MÉRIDIONAL.

| Noordzee (Holland) | Nordsee (Holland) | | Middellandsche Zee | Mediterranean Sea |
| North Sea | Mer du Nord | | Mittelmeer | Mer Méditerranée |

Fig. 7

Platte grond
Plan   Ebene
Plan

Fig. 8

Amsterdamsche voeten
van 11 duim

PINASSCHIP
PINASSE   1625   PINNACE
PINASSE

Early square sails in the North had been wool or even leather, but were replaced by lighter flax, which was easier to manage and repair; more and smaller sails divided among different masts also made ship-handling easier. Side rudders or "steer-boards" (mounted by the Norsemen on the right side of the stern, thus the term "starboard") gave way to the center-mounted stern rudder that could turn a vessel faster; it was controlled by a helmsman inside the ship using a whip staff attached to the tiller below. Because their builders lacked the saw, the hulls of early northern ships were clinker-built, with overlapping timbers laid on a relatively light frame. In the South, carvel or flush-built techniques prevailed, and though the resulting vessels were leakier and demanded heavier framing, they could also be built to greater length.

Meant for commerce, most discovery vessels were only lightly armed with a few cannon and small ordnance like the falconet and the swivel gun. Gunpowder was hard to keep dry in leaky holds, and when it was dry posed a constant threat of explosion. Swords and pikes were still familiar weapons to seamen into the eighteenth century. The most fearsome weapon we know that Cabot carried in the *Mathew* was a crossbow.

The ships of the age of discovery were not custom-built for long ocean voyages, but rather were adapted to that purpose from existing designs. The strength and seaworthiness characteristic of knockabout merchantmen also met the needs of the new discovery enterprises, which were largely commercial in nature. (Thomas Fisher Rare Book Library, University of Toronto)

strait it guards, which would have looked like the passage to China he had set out to find. But, perhaps because of icebergs, which are still common in June in these waters, he turned away and probably sought shelter in Quirpon Harbor or Griquet Harbor, which sits just at 51 degrees 33 minutes north, exactly the same latitude as Dursey Head, where he had set off. He landed and formally staked the claim of England, unfurling the cross of St. George for Henry and the cross of St. Mark in memory of his hometown, Venice. He may have taken a shore party to the top of Cape Dégrat and surveyed the ice-plugged passage to the west and the vast landscape behind him, which certainly must have looked like the mainland of something.

The later accounts of Raimondo Soncino (the Milanese envoy in London) and of John Day (an English merchant who was in Bristol when Cabot returned from his voyage and whose famous letter to Columbus about it contained precise navigational details that pretty well scotched the old Canadian claims that Cabot's landfall was on Cape Breton Island) tell us that Cabot's party saw signs of human habitation—fishnets and snares, and perhaps part of a shuttle used in the weaving of nets—but no natives themselves. When in 1501 the Portuguese Gaspar Corte Real kidnapped fifty-seven Beothuk Indians and took them to Lisbon, they supposedly displayed an Italian sword and a pair of earrings, which, if they existed, could have been Cabot's. Morison plots a subsequent itinerary that took Cabot the length of the east coast of Newfoundland, across White and Notre Dame bays to Fogo Island, then south past Cape Freels, around Cape Bonavista; perhaps he looked into Trinity Bay, noted St. John's Harbor, passed Capes Broyle and Spear and finally moved on to Cape Race, Newfoundland's southeastern corner and the landmark that steamers on the Great Circle route from England and France try to make before changing course toward Halifax, Boston or New York. (When the *Titanic* sent out her famous "CQD" distress signal on April 15, 1912, Cape Race was the reference point.) Cabot then proceeded around Cape St. Mary's toward Placentia Bay and the Burin Peninsula, beyond which lies the broad southern passage (later called the Cabot Strait) into the Gulf of St. Lawrence but which would have looked to Cabot like the strait to China. But (if any of this conjecture is correct) with just one ship and with plans to return soon with a larger expedition, Cabot at this point put *Mathew* about and retraced his path to Cape Dégrat, and then on July 20 set sail for Bristol. *Mathew* made a speedy fifteen-day passage, although Cabot missed England and first raised Ushant Island off the western tip of Brittany. He made Bristol on August 6.

Cabot proceeded directly overland to London, where he reported to Henry his discovery of a new island; the map that he showed the king has been lost. Henry named it, straightforwardly, New Isle, but within a few years Newfoundland seems to have taken hold. Henry evidently was pleased enough, although still not especially generous. He presented Cabot with ten pounds ("To

Fifteenth-century English ships depicted in the Hastings Manuscript foretold the rise of the maritime ambition that would challenge Spain for supremacy in the Old world and the New. (Pierpont Morgan Library)

Sebastian Cabot may have accompanied his famous father, John, on his 1497 voyage from Bristol to Newfoundland. Throughout his life, he skillfully promoted himself on the basis of it.(Massachusetts Historical Society)

hym that founde the new Isle") and in December added an annuity of twenty pounds, payable out of the Bristol customs.

For the next voyage Henry actually came across with a ship (we do not know its name), crew and provisions, and with new letters-patent dated February 3, 1498, which authorized Cabot to charter up to six English ships of up to two hundred tons and instructed him to sail "until he reaches an island which he calls Cipango [Japan], situated in the equinoctial region where he thinks all the spices of the world have their origin, as well as the jewels." He was also to set up a colony, or a trading post, whereby the merchants of London could better insert themselves into the spice trade. Actually, it was the merchants of Bristol who outfitted four ships with trading truck for the natives, and whose investment measured their substantial expectation that this time the Genoese Cabot would bring home the gold. It was not a profitable gamble for them, and no ticket at all for poor Cabot. Not only did Cabot not bring home spices and precious stones but neither he nor four of his five ships returned at all. (The fifth had turned back just past Ireland.) They set forth at the beginning of May 1498 and were never heard from again. Henry paid the pension for the last time to Cabot's widow following Michaelmas 1498, and then presumably closed the book on Cabot and on North America. It would be nearly ninety years before his granddaughter, Elizabeth I, reopened it.

Cabot had bad luck: a niggardly backer and some fatal accident at sea. No greater, more determined patron subsequently appeared and no one followed up Cabot's discovery. No American treasure flowed into England's exchequer, except what she could steal from Spain. The English had to keep getting their cloves and cinnamon the old way, via the Levant. Newfoundland remained as ever shrouded in fog and ice, her wretched Skraelings as ever undisturbed. After Cabot, the fishermen continued their businesslike voyages to the great cod fishery of the Grand Banks, just as they may have done even before him, but they went that far only because they had to and certainly not for the romance or adventure of any possible "discovery." No one guessed what actually lay behind the Newfoundland barrier across the great gulf, and even after Cartier told them, no one could yet do very much about it. The lands to the south, between Newfoundland and the Spanish possessions, remained a blur even after Verrazzano's survey of that long coastline, for what was still missing was any sound understanding of what exactly it was the coastline of. If you sailed west for thirty days or so, it was getting to be a well-established fact that you would hit land. Except for the cod fishermen, who didn't much care what it was, pretty much everyone else who made the trip saw in that land what they had set out to seek, and that was Asia.

All the longitudes were impossibly wrong, but everyone then believed Asia was much larger than it is. Even as the empirical evidence began to pile up and the observed characteristics of the

The traverse board helped a navigator to plot the true forward motion of a vessel when tacking or beating to windward; the information then had to be transferred to a chart at every observation of the sun. (National Archives of Canada)

place people kept bumping into seemed less and less Asia-like, "America" intrigued Europeans only as an obstacle to be gotten around, not as an opportunity to be exploited. Until that perception changed (and it did not change for the northern Europeans until late in the sixteenth century at the time of the Roanoke voyages), no progress would be made, because if Asia was what you really wanted, then Columbus had it all wrong. Sailing west to get east was not the best way to do it. For if you discounted the Strait of Magellan, which is really just a minor shortcut around the far bottom end of the American landmass, and the Northwest Passage, which proved illusory for everything except a supertanker in 1969 trying to prove the point and prowling nuclear submarines, then there is no strait worthy of the name. At least there was none until the Americans, centuries later, dug one with dynamite and steam shovels through the jungles of Panama.

Christopher Columbus can be said to have discovered America not because what he at first found (the Bahamas, Hispaniola and Cuba) was any closer to the goal of Asia than what Cabot found (Newfoundland and who knows what else on his ill-fated second voyage), but because the news he brought back fairly electrified his Iberian patrons and galvanized them to take quick and full advantage of real estate that seemed very much up for grabs. In another respect too, Columbus was lucky where Cabot was not. The southern territories Columbus came across happened to be home to some relatively advanced native populations—not so advanced as to be able to offer significant military resistance to the conquerors, but advanced enough to have found and stockpiled vast hordes of gold and silver and precious stones ("treasure" just as the Europeans understood it). But for this very tangible jackpot, the first jolt that Columbus's discovery gave to the Spanish may well have dissipated fast enough, and southern and central America soon came to seem just as boring to the Spanish as northern America long was to seem to the French and the English.

Columbus's bibliography is enormous—it dwarfs Cabot's. We have his journal, we think we know what he looked like. He remains the subject of lively scholarly discourse: a Columbus Landfall Symposium convened at The Johns Hopkins University, in Baltimore, in October 1989 to argue about where exactly the Genoese first set foot to American sand. He indubitably laid the first foundations of the Spanish empire in the New World, which lasted more than four centuries until the Spanish-American War of 1898. He became a great man in his own time, a legend ever after. Schoolchildren who know no other history know his name and the names of his three ships and can tell you what he did and when. The place where he landed on the Bahamian island of Guanahani, as the natives called it (later to be known as Watlings Island), is marked with a cross to this day. Indeed, you can if you like follow his trail all through the Caribbean; Samuel Eliot Morison and an expedition from Harvard University did just that with painstaking thoroughness in 1940.

William Bourne's Regiment for the Sea *(1574) was one of many books on navigation published for a maritime audience that grew as the horizons of the world expanded. (British Museum)*

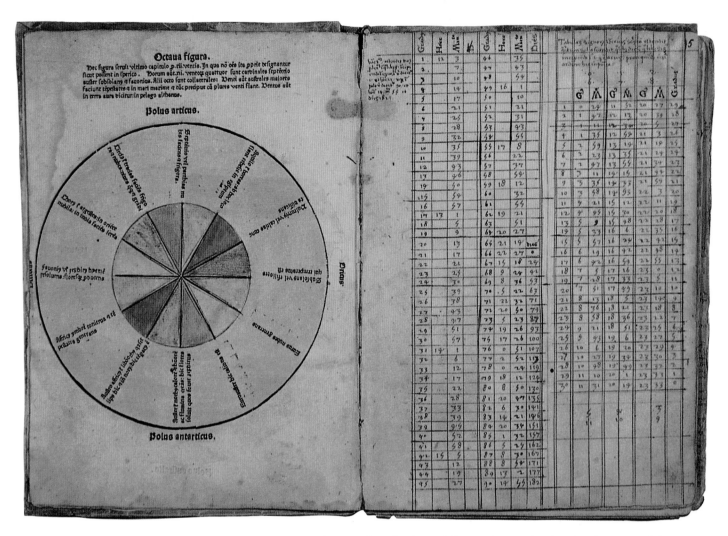

Columbus's logbook reveals him as a meticulous record-keeper and a captain keenly conscious of the momentous nature of what he was about. (Biblioteca Colombina, Seville)

But the truth of what followed, in the lands that would one day be the United States and Canada, is not best approached through the roamings of the great "Admiral of the Ocean Sea," which was the title bestowed on him by Ferdinand and Isabella. What happened to the north in the wake of Cabot's Columbus-like discovery of Newfoundland—nothing much—was utterly unlike the hell-for-leather Spanish gobbling up of a whole continent that occurred in the decades shortly after 1492. For the first 150 years, the results of discovery in the north were not much to behold: even by the middle of the seventeenth century, there was little more than a string of seaboard settlements dependent on trade with the motherland, backed against a vast wilderness filled with wild men untamed and unreconciled to European permanence in their land. Everything in the North took longer to get going, and, because there were not one but two colonizing powers, history there took on added contentiousness. For a century and a half France and England sparred continuously and then finally in the 1750s fought to the finish over Canada, leaving that country with its divided cultural inheritance. The irony is that after the slow start, the lands sprung from the northern discoveries would run faster and farther than anything in what would come to be called Latin America. The United States and Canada modernized apace with northern Europe. From human intelligence and nature's crude bounty they created wealth, if not well-being, such as the world had not seen before and that made the riches of the Inca and Aztec seem primitive and puny. The northerners pioneered and practiced theories of politics that gave them stable governments and relatively enfranchised and contented citizenries. No southern American country enjoyed a remotely similar history.

Quirpon Island sits at 51 degrees 35–38 minutes north latitude and guards the northernmost water route into the heart of Canada (if you do not count the circuitous arctic path through the Hudson Strait to remote Hudson Bay). It overlooks the Strait of Belle Isle between Newfoundland and Labrador, and is separated from the mainland of Newfoundland by a narrow but navigable strait. To reach it you can in winter walk across the ice; in summer you must find a local fisherman willing to ferry you over in his boat. There is no formal public access, no visitors' facilities, no cars, in fact no roads, no paths, no people.

The island is an undulating, boggy place, about five miles north to south and two miles across, its surface a soft cushion of peat, with a score of small ponds breaking up the moor. It could be Scotland or Yorkshire, Somerset or Devon. Gorse, heather and lichens carpet it; scrub elder fill the hollows; there are no proper

The view from Cape Dégrat toward Cape Bauld at the northern tip of Newfoundland. The 500-foot promontory of Cape Dégrat was probably the site of John Cabot's North American landfall on June 24, 1497. (Dudley Witney)

trees. Uninhabited but for the lighthouse keeper on Cape Bauld at the far north end, it is a lonely place but not an unwelcoming one. Stick to the high ground and you'll be all right, I was told after announcing my desire to go to the summit of Cape Dégrat. It is a one-and-a-half-hour walk across an open landscape with no trails and a stiff climb toward the end. To Cape Bauld, it is flatter but twice as far. But the altitude here is important, for only from some height can you take in the sea: the Atlantic and the Strait of Belle Isle. At the top, a small (three by three inches) brass plate, embedded in the rock and inscribed by the Canadian geodetic survey, pinpoints the coordinates and the elevation and reminds you that Canada's is now a thoroughly marked and measured coastline. Due north, you can see Belle Isle across the strait; due east over the curve of the earth lie County Cork, Bristol and London; due south, the long reach of Newfoundland's Great Northern Peninsula; due west, the immensity of Labrador and three thousand miles of Canada reaching to the Pacific; straight down, the rock-bound shore that has tested the mettle of mariners for centuries and claimed more than a few of them. At the top, after the climb, as I sat in the lee of an outcropping, out of the wind, it occurred to me that if one were just starting out in a new world, perhaps a little unsure of whether to go or to stay, then

Columbus's Santa Maria *is the most famous of all the ships of discovery, shown here longitudinally and in cross-section. She was not, however, discovery's most successful vessel, fetching up as she did on a Caribbean reef and never returning to Spain. (Thomas Fisher Rare Book Library, University of Toronto)*

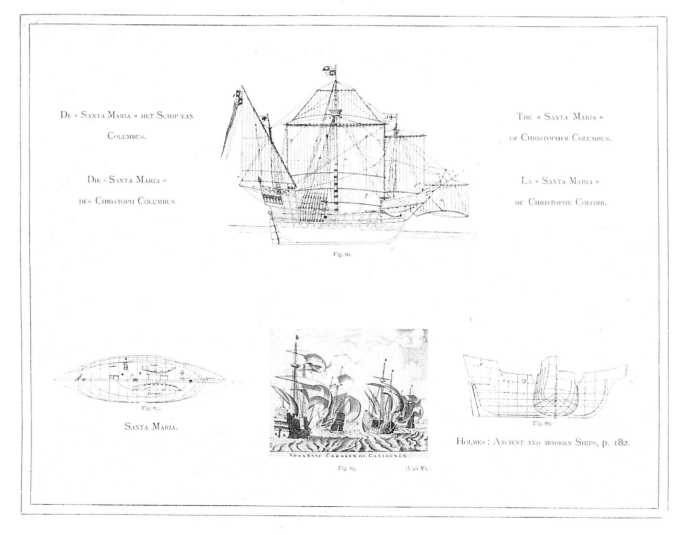

Model of the Santa Maria, *fully rigged. (Science Museum, London)*

this wouldn't be a bad spot to pause and sort things out. It would have been at least as good as L'Anse aux Meadows and, as an island, a good deal safer.

The Quirpon fisherman who cheerfully took me over had fished these waters all his life, and his father and grandfather before him. In the summer of 1989, he complained of not many fish close in where you could get at them; as every year passes the great factory ships of foreign countries progressively suck Canada's waters clean. He seemed a proud man, after the measured manner of people who live close to nature and make their livings from it: proud of his old wooden boat, proud of his new fiberglass boat, proud of his new house built with his own hands, and no doubt proud of all the latest electronic home entertainment gear that festooned his living room. Pampers swathed the baby, but the tea and cakes we had from his shy and gracious wife were warm and genuine. He also seemed proud of the old dock house where I first found him out over the water's edge, built by his forebears

Christopher Columbus

HIS NAME IS THE NAME OF DOZENS OF CITIES AND TOWNS all across the United States; it accounts for the "C" in Washington, D.C. (District of Columbia); it was the name of the greatest world's fair ever held, which marked the 400th anniversary of his great discovery: the World's Columbian Exposition of 1893. Small schoolchildren know his name, the date of his discovery, and the names of the three ships he sailed west to reach the East: *Niña, Pinta* and *Santa Maria.* Columbus Day is an American national holiday.

His greatest biographer, Samuel Eliot Morison, relates the story of Saint Christopher, Columbus's patron saint, to illustrate the religious core of the great man's character. It is the tale of an enormous pagan in search of Christ who, because he cannot fast and pray, carries poor travelers on his shoulders across a river where there is no bridge, in the hope that the Lord will come to him. One night, he bears across a child whose weight mysteriously increases until Christopher nearly breaks beneath it. "Marvel not, Christopher," his passenger explains, "for thou hast borne upon thy back the whole world and Him who created it. I am the Christ whom thou servest in doing good." Columbus certainly knew the story, and saw it as part of his burden to carry the Gospel of Christ across the ocean sea to the heathen.

He was born in Genoa, probably in the late summer or early fall of 1451, was raised up among the seafaring folk and traditions of the Ligurian Republic, and was at sea by the early 1470s. Finding his way to Portugal, then the center of seaborne discovery, he conceived his *Empresa de las Indias,* his "Enterprise of the Indies," and from 1483 onward he spent every ounce of strength finding a patron for an expedition to prove the existence of a short sea route to Asia—not by sailing south and east around Africa, but west across the Atlantic. Columbus's speculation about the width of the Asian continent and thus of the narrowness of the Atlantic was wildly inaccurate: he thought it was 2400 nautical miles from the Canaries to Japan; it is actually 10,000. No ship of that age could have made such a voyage nonstop, and had the true distance been known, no backer would ever have taken the risk. As it was, his search for a patron took years, and when Ferdinand and Isabella of Spain at last assented to his terms, it was likely with skepticism that this visionary Genoese would ever live to tell his tale. But no one bet on America, the vast continental landmass that lay astride the path to China.

Columbus set sail from Spain in August 1492, weighed anchor from the Canaries on September 6, and so bade farewell to the known world. By dead reckoning and rudimentary celestial navigation, he guided his little fleet westward along the twenty-eighth parallel to where he thought Japan would lie. Instead, at 2:00 A.M. on October 12, a lookout on the *Pinta* sighted the white moonlit cliffs of the Bahamian island of Guanahani. The next morning, Columbus made his first and greatest landing on a New World shore and called it, in keeping with his faith, *San Salvador*—Holy Savior. To his patrons, he prophesied with amazing clarity that his discovery would turn countless people to the holy faith, to say nothing of the material benefits, "since not only Spain but all Christians will hence have refreshment and profit."

Columbus, who was dubbed "Admiral of the Ocean Sea" by his patrons Ferdinand and Isabella, boasted a personal coat of arms festooned with anchors and islands to prove it. (Palazzo Tursi, Genoa)

from local timber, part boat house, part toolshed, part sawmill and woodworking shop, it housed nets, buoys, motors, oil drums—yes, even a few fish.

He was also proud of his harbor. On learning that I had come to his far end of the world not just to see well-known L'Anse aux Meadows with its fancy visitors' center and park service paraphernalia but also to climb the promontory that Cabot and Cartier probably had climbed, he showed obvious pleasure. It wasn't, you see, that he had particularly strong feelings about either Cabot or Cartier, though he knew it was part of local lore that the early explorers had been there. But he did have something to say about L'Anse aux Meadows. Now Helge Ingstad was a good archaeologist, and the Norse nails he found do not lie. But my fisherman tempered such scientific conclusiveness with a mariner's practical question. If the Vikings, he asked, were the great sailors they must have been to come this far in boats not a lot bigger than his own new fiberglass number, and assuming they knew about Quirpon Harbor, then why would they have chosen the disadvantageous spot they did for their settlement? Épaves Bay is more a broad curving beach than a proper harbor; it offers no shelter at all from the northwest, and its waters are, well, nothing but shoals. True, the hinterland is nice, but it was nice for the Skraelings too. Why not Quirpon instead, with its fortress of an anchorage and its great sheltering island, close enough to the main to get what you needed but eminently defendable against those you wanted to keep out?

L'Anse aux Meadows on Newfoundland's Épaves Bay is but a stone's throw away from Cape Dégrat. (Dudley Witney)

Remember how, centuries later, the adventurers at both Roanoke and Jamestown sought out islands for their first tender settlements, and neither of those islands was nearly as impressive as Quirpon. My host wondered, reasonably enough it seemed to me, whether the Vikings hadn't had another place on Quirpon, in addition to L'Anse aux Meadows, one that just awaits digging up. And if they had, well then wouldn't it be a bit of an embarrassment to the Parks Canada and the UNESCO people, "With all that money spent over there at L'Anse aux Meadows?"

I suppose it would be, which means that Quirpon isn't going to become a World Heritage Site anytime soon. L'Anse aux Meadows has that distinction, along with all the fancy interpretation. Quirpon has only Cabot, Cartier, Leif and Thorfinn and crew. But it has this colossal advantage: it is today a living place and not a museum. To my friend, the fisherman, it's a fine island, a great place for kids to romp around in the summer and for grown-up kids to go snowmobiling in the winter, a great place for picking bakeapples and blueberries and for sighting the occasional eagle. Its coves are still good places to fish: from atop Cape Dégrat you can look down to the sea onto the red floats that mark where the fishermen will haul their nets. My friend will sell his catch to a commercial buyer and keep something for his family. The best cod, he says, are still sun-dried on the wooden racks that the tourist brochures trot out as something of a Newfoundland totem. But there's little left of it. Over in St. Anthony, you buy your cod fresh-frozen, just like the rest of the world, at The Viking Mall.

What to trust: the artifacts of the archaeologist or the instincts of the fisherman? Whichever, the coincidence of historical events is striking, and this is the place of all discovery places where it is best observed. This tiny top tip of Newfoundland comprehends the entire five-hundred-year history of Canadian discovery from the end of the tenth century to the middle of the sixteenth, and there is something about the character of Newfoundland today that reflects back on that past. None of them—the Vikings, Cabot, or Cartier—stayed very long. They all made brief excursions here while they were really looking for something else; they happened onto Newfoundland and stumbled into Canada. In this tentativeness there was a certain foreshadowing. Newfoundland developed, to the limited extent it did, obedient to the thing that had interested Europeans in it in the first place: fishing. All sides face the sea and the banks of cod, the whales, the seals. For years the economy was extractive and fundamentally colonial. To this day its little coastal villages strike the eye of the non-Newfoundlander with the queer randomness of their spatial arrangements. No hint of section-line, rectangular survey, seigneurial boundary defines them, but only a colorful hugger-mugger of houses, churches, stores, fish factories strewn over some rocky headland or tucked into some sheltering cove. There seems no pattern, and there is none. There is only the draw of the sea.

Until very recent times—the 1950s and 1960s—land transpor-

Columbus's long letters and sea journals set him apart as one of history's most self-conscious discoverers. (Archivo General de Simancas)

THE VOYAGES OF DISCOVERY WERE UNDERTAKEN FOR profit, and outfitting of the ships that did the job reflected rudimentary but not unusual standards of provision. Sailors usually supplied their own clothing: baggy trousers, long woolen stockings and an overgown made of coarse serge with a hood to protect against the elements. Changes of clothing were not the rule, and the Royal Navy did not undertake to provide extras until the seventeenth century. Hammocks had not yet come widely into use, and men slept wherever they could find a spot; bunks were for officers only.

Pickled pork and beef, and "bisket" or hardtack (baked on shore and almost always filled with mold and maggots by the end of a long voyage) were a seaman's staple diet. On voyages to North America, sailors might look forward to supplementing their usual fare with dried cod, fresh fish and wild game. Beer and cider, not water (except in the last resort when the beer went sour), slaked the thirst. Cooking was done in the "cook-box," usually located forward in the ship and sheltered by the forecastle, and consisted of a simple iron box on a bed of sand. The term "galley" did not come into use until the eighteenth century.

Sanitation was primitive, ballast and bilge water commonly polluted with garbage, human waste, dead mice and rats. The tradition of the sea prescribed the presence on board of one or more cats to keep down the rodents; there was nothing to keep down the roaches, lice and assorted other shipboard vermin.

For their labors and their risk (which was considerable), merchant sailors in the sixteenth century earned a wage of perhaps five to ten shillings a month, while a "portage" or share system was the rule in the fishing fleets. The age of discovery was also an age of religion, and the spiritual needs of sailors were met perhaps more adequately than their physical ones, which cost the owners money. Part of the kit of most English captains was the 1549 *Book of Common Prayer*, and many read morning and evening services to their crews who, whether devout or merely superstitious, were quick to see themselves in the words of Psalm 107: "They that go down to the sea in ships: and occupy their business in great waters; These men see the works of the Lord and his wonders in the deep."

The Sailor's Life

The association of travel with ships and the sea is an ancient one. Here, Marco Polo, the world's greatest explorer before the age of American discovery, meets the waterfront. (Metropolitan Toronto Reference Library)

For every sailor who actually went to sea, dozens more worked on the shore, on the quays and in the shipyards. Early wood-cuts depicted them. (Metropolitan Toronto Reference Library)

tation in Newfoundland was primitive. With immense difficulty, one trans-island railway was punched through from St. John's to Port-aux-Basques in the late nineteenth century, and the age of the automobile and the paved highway also came late here. The outport villages all around the coast had no overland connection with one another or with St. John's. If you lived in Quirpon and wanted to visit St. Anthony, it was easiest to go by boat. Your fish went out by boat, and your supplies (what you couldn't make or grow for yourself) came in by boat. Today, however, the roads have come, and various governments have pursued resettlement programs aimed to entice poor fisherfolk away from their quaint but impoverished isolation into larger towns and more progressive ways of life. Thus the outports are much diminished, though a few remain along the south coast, reached only by steamer from Port-aux-Basques or Terrenceville: La Poile, Grand Bruit, Grey River, François, McCallum, Gaultois, Rencontre East. Their disappearance and the neglect of the fishery are today the source of much local bitterness toward the federal government in Ottawa. Partly it is the lament of a dying breed of independent folk who acknowledged no masters but the sea and the market for cod, and whose voice can be heard in the songs of Quirpon fisherman Wayne Bartlett and his Men of the Northern Peninsula, Sim Savory, Wilfred Sullivan and Hughie Poole:

> *The fisherman's a dying breed*
> *No one can take his place,*
> *And pretty soon that memory will all but be erased.*
> *No one can live on promises, no one can live on need,*
> *For God's sake smarten up, my friend, and help the dying*
>   *breed.*

Among the cassettes for sale in the grocery-confectionery stores that dot the island, you will find a Newfoundland popular song whose theme returns to the origins of this Britishness, if not quite all the way back to Cabot. Recorded by a group that smartly calls itself The Saltwater Cowboys, the song title, "Heaven by Sea," refers to the reported last words of Sir Humphrey Gilbert, favorite of Elizabeth I before Sir Walter Raleigh, and who sailed on her warrant in 1583 to colonize the New World for England. Four ships made the crossing (*Delight, Golden Hinde, Swallow* and *Squirrel*), and on August 5 Gilbert formally laid claim to Newfoundland in St. John's harbor before an assembled fleet of Spanish, Portuguese, French and English fishermen. From that moment, Newfoundland dates the beginning of its cherished British association.

Unlike many of his contemporaries, Gilbert was not a fortune-hunter alone. He had in mind genuine long-term colonization and saw it hopefully as an outlet for unemployment at home and the beginning of a happier social order. He did not live to see the

QVID NON:

VIRGINIA

Adventurer, West Countryman and half-brother of Sir Walter Raleigh, Sir Humphrey Gilbert obtained the first patent from Elizabeth I for colonization of northern North America. He founded the colony of Newfoundland at St. John's harbor in 1583, but was drowned when the pinnace Squirrel *foundered on the return voyage. His motto was Quid Non? (Why not?). (British Museum)*

venture through and was saved by an early death from probable failure. On the return voyage, somewhere north of the Azores, the fleet confronted terrible seas, and the little *Squirrel*, a pinnace of just ten tons, was swamped by a huge following sea and disappeared without a trace with Sir Humphrey aboard. Shortly before the disaster, the *Golden Hinde* had come within hailing distance, and her master, Edward Hayes, reported Gilbert sitting abaft reading a book and calling across the tempest: "We are as neere to heaven by sea as by land."

Newfoundland's future did not develop along neat utopian lines, and when Elizabeth reassigned Gilbert's grant to his half-brother and her next great favorite, Sir Walter Raleigh, it was henceforth Virginia and not Newfoundland that became the focus of her country's colonizing enthusiasm. But even as brief as his story was, Sir Humphrey spent more time in North America than Sir Walter ever did (none at all), and his name is remembered by the locals (it was my friend the fisherman who told me there was a song about Gilbert), by The Saltwater Cowboys, and even by the Canadian Coast Guard, which named an icebreaker after him. As a founding father, Sir Humphrey seems well fitted for Newfoundland, for a very tentative founding it was, of a place whose future was ever touch-and-go, and whose relationship with its big Canadian brother with whom it finally joined up has ever been an uneasy one.

Navigation

GETTING THERE AND BACK TO TELL ABOUT IT—THE JOB OF discovery—was something that mariners by the sixteenth century were reasonably well-equipped to do. Geographical knowledge and navigational know-how may have been woefully lacking by modern standards, but what more than compensated was the unwavering conviction that lands across the sea did in fact exist, and the courage and technical skill that could sail small ships to them.

The most valued navigational instrument to any captain was his compass which, two centuries before the age of discovery began, had been developed to the point where lengthy voyages out of the sight of land, or in conditions of low visibility, could be undertaken without getting lost. With a compass, it was possible to follow a course within several degrees, and beginning in the fourteenth century it was used to compile "portolans," or pilotage manuals for much of Europe's coastline.

The instruments available for celestial observation were primitive and less reliable, and were a good example of a theory that had been worked out while the mechanics still lagged behind. But in the hands of an already practiced mariner these instruments did yield successful if not spectacularly accurate results. To determine a ship's position north or south relative to the equator, it was necessary to measure the angle between the horizon and either the North Star or the noonday sun and then, through the use of sets of tables worked out by astronomers on land, calculate degrees of latitude.

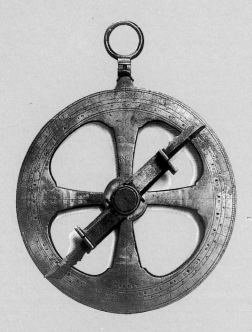

The astrolabe had been used for 2000 years to measure the altitude of the sun, but it was designed by and for landsmen. Enormous errors could result when it was employed on the rolling deck of a small ship. (National Archives of Canada)

Early mariners relied on navigational guides as well as instruments. This is a page from a 1545 publication. (Bibliothèque Nationale, Paris)

The astrolabe had been in use for 2000 years in measuring the altitude of the sun, but enormous errors resulted when it was used on the rolling deck of a small ship. The answer to the problem was the cross-staff—a stick fitted at right angles with sliding crosspieces which, when aligned with the sun at one end and the horizon on the other, yielded the sun's altitude. "Latitude sailing" was the result, whereby the navigator first set a course north or south until he fixed the supposed latitude of his destination by celestial observation, and then sailed straight (or as straight as he could) east or west until he sighted land.

Proper calculation of longitude awaited development and application of accurate clocks in the eighteenth century. Until then, time at sea was kept by the sandglass, which had to be turned every hour or half-hour for the length of the voyage. Many mariners conservatively relied on "dead reckoning," which required measuring the speed of the vessel and then estimating distance traveled along a particular compass bearing. When contrary winds made tacking, or zig-zagging necessary, the traverse board helped plot forward motion.

Cartier's
Conquest

ABOUT HALFWAY UP THE EAST COAST OF CANADA'S
Cape Breton Island along the Cabot Trail, there is a lovely place
to have lunch. With its gables and chimneys and broad lawns and
brightly painted Adirondack chairs, the Keltic Lodge is one of those
old-fashioned-feeling resorts where you can still take large meals
and read quietly in front of the fire. It opened as a hotel in 1940
on property that had once belonged to Henry and Julia Corson,
rich Americans from Akron, Ohio, and friends of Alexander
Graham Bell, who lived not far away in Baddeck. After fifty years
it continues to operate with a certain flair for the old traditions.
It is run by the Nova Scotia Department of Tourism and Culture,
which proudly touts its golf, tennis, heated pool, gourmet
cuisine—and "spectacular vistas of land, sea, and sky." With more
than just the name, it celebrates the Scottish heritage of the region:
kilted footmen greet your car at the gate and usher you into a
lobby carpeted with tartan.

If you are a Scot and want the full treatment, however, the town
of South Gut St. Ann's (population sixty-four) offers the Nova
Scotia Gaelic College of Arts and Culture, the only Gaelic college
in North America, where you can learn the language, play the
pipes, try the highland dance and song, weave the tartan. There
is a Great Hall of the Clans, and elaborate entertainments, including the Ceilidh and the Gaelic Mod, a seven-day festival of Gaelic
culture held every August. Nearby at Englishtown, a fishing village
on St. Ann's Bay, you can view the grave of Angus MacAskill,
"The Cape Breton Giant," who at seven-foot-nine and 425 pounds
won acclaim as the strongman of P. T. Barnum's circus in the nineteenth century. Except for the cars driving on the right-hand side
of the road, there is little in the natural or the human scene to
say that you are in Canada and not the Highlands. The Scottish
character of Cape Breton Island dates to the Highland Clearances

*Theophile Hamel's portrait of Jacques
Cartier depicts a seaborne adventurer.
Cartier's voyages to Canada in the 1530s
revealed something of what the land
promised, but it would be decades before
permanent white settlement followed.
(National Archives of Canada)*

of the 1740s and the Great Migration that began to scatter the Scots far and wide across the globe. Nowhere could they have found a place that looked more like home than this, and nowhere has their transplantation been more reverently enshrined than in this province of Nova Scotia—Latin for New Scotland.

But from a window table in the dining room of the Keltic Lodge, the "spectacular vista" looks out onto a coast with a history older than the history of the Scots in Nova Scotia. After lunch, follow the trail out onto the peninsula called Middle Head, and from there behold the two broad bays of Ingonish. Ingonish today is a popular resort destination, its little communities of Ingonish, Ingonish Centre, Ingonish Beach, South Ingonish Harbour and Ingonish Ferry attracting a wide spectrum of visitors to the lovely scenery, abundant outdoor recreation, and accommodation ranging from the grandeur of the Keltic Lodge to primitive campgrounds. Not everyone, though, caters to the tourists.

It was not, after all, scenery that first brought white men to these shores. It was the fish. While somewhat diminished, commercial fishing still plays its part in the local economy and contributes to the charm that also makes the tourists come. "A good spot to buy fish and lobster directly from the fishermen," reports *Nova Scotia: The Doers and Dreamers Guide* (the official province guidebook) of Neil Harbour, just north of Ingonish. The same could be said of a score of other little places around this rugged island—almost a peninsula—that juts up fiercely between the Gulf of St. Lawrence and the Atlantic and points the way toward Newfoundland. And the promoters are right. To vacationers from less maritime milieux (which means almost anywhere else than Cape Breton Island), used to their fish frozen in the supermarket, there is genuine romance in the discovery of men who still, every day, go down to the sea in little boats and cast their nets on the broad waters.

For the men who do, it is a living, and from the promontory of Middle Head you can watch their boats. They do not want for predecessors. Four hundred and sixty-five years ago, the boats of Portuguese fishermen filled the bays of Ingonish, aiming to fill their holds with these waters' richest resource, codfish. João Alvares Fagundes is the Portuguese whose name is most often associated with this place. It is known that on May 22, 1521, he received a grant from the king of Portugal to lands in this region, and that with colonist-fishermen recruited in his native Minho in Portugal and in the Azores, he sailed to Cape Breton Island sometime between 1521 and 1525. He believed that it would be more efficient if, instead of having to return to Portugal with every cargo of cod, his fishermen first could cure their catch ashore. Fagundes stayed a year, perhaps a year and a half, at Ingonish's protected harbors. But he had no help from home and no real hope, and probably no intention, of sustaining a colony in North America. The Indians grew hostile, and Breton fishermen who also knew these waters did what they could to discourage his intrusion. It was only a brief episode, but the argument can be made

French maritime technology blended northern and southern influences. This model of a fifteenth-century ship betrays the stubby build of the coastal merchantman, which was remarkably well suited for the trans-Atlantic passage and the exploration of uncharted shorelines far from home. (Maritiem Museum Prins Hendrik)

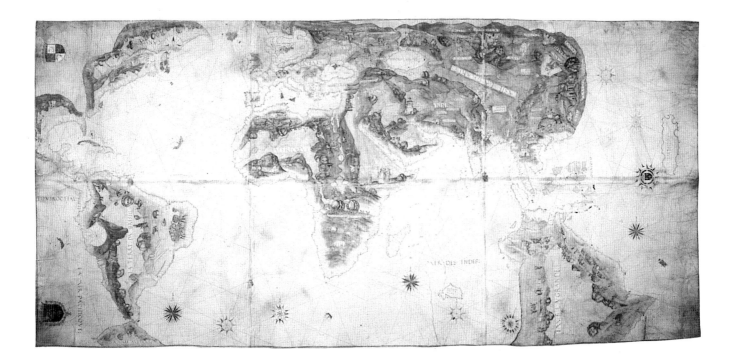

that this was the site of the first European attempt at settlement on the North American mainland since the Vikings quit the coast of Newfoundland five hundred years earlier.

From Middle Head you have a good vantage not just of the Ingonish harbors where Fagundes cured his cod and careened his boats but also out across the broad expanse of waters that in the sixteenth century, and probably earlier, drew the fishing fleets of many nations. Tierra de Bacallaos, Codfish Land, the Portuguese called it. They meant, most particularly, the Avalon Peninsula on the southeast corner of Newfoundland and the Grand Banks that reach out over three hundred miles southeast from Cape Race and that are home to vast shoals of voracious cod. But the seas all around that enormous island and the adjacent mainland coastlines abound with sea life and together form the greatest commercial fishery the world has ever known. First to take advantage of it were the stout little tubs, with illiterate crews whose names went unrecorded, from the Basque country of France and Spain, and from Portugal and England, who harvested the cod for the fish-hungry markets of a still very Catholic Europe.

Of all sailors, fishermen least like to advertise their whereabouts to the competition, and so the history of their voyages is shrouded in conjecture. What is not conjecture is that it was the sea and its bounty that interested them, not the land with its rocks and trees and sullen natives. Yet a look at any modern map will hint at what these simple cod fishermen were at least getting close to. Cape Breton Island and Newfoundland sit astride the entrance to an enormous gulf, and the gulf leads on to a watery path of rivers and lakes that reaches far into the heart of the North American continent. The cod fishermen did not know this; they were practical men bent on making a living to enjoy back home, and not explorers

The Harleian map was probably made about 1544 by Pierre Desceliers or Jean Rotz. Cartier's discoveries on his 1535 voyage along the St. Lawrence are depicted on the map. (Courtesy of the Trustees of the British Library)

charged to lay hold of new lands for proud monarchs or to find new routes to the riches of the Indies. But along with their homely cod, they brought home, surely, all manner of tales as well.

By the score, their ships commuted across the Atlantic once, sometimes twice a year, years before any settlement that could be called a proper colony existed on the far side. Because of their uncounted voyages, fish found its way onto Europe's tables. But not fish alone: North America quietly, subtly, in ways it is impossible to document or measure, found its way into Europe's consciousness. Because of the fishermen, Europeans grew in confidence that the Atlantic was as much pathway as barrier. Their voyages demonstrated that it could be crossed, if not without considerable peril (long-distance travel almost anywhere in the sixteenth century was perilous), then at least with great regularity and in the course of conducting ordinary business and with resources that could be mustered without resort to great public enterprises. While the technology of navigation remained crude at its best, and while many mariners no doubt did not take advantage of the best available tools, even so the shape of remote coastlines was memorized, the patterns of wind and current and ocean swell near land became familiar. Habits took hold, habits that rendered the Atlantic passage into a routinely doable task, whether the object was cod or colonies.

Of all the Europeans who frequented the cod fisheries in the sixteenth and seventeenth centuries, the Basques from Bayonne, St. Jean de Luz, Ciboure, Amuix, San Sebastian, Montrico and Ea offer the best study of seafarers in pure culture: largely unconcerned with the land, they nevertheless left their markings all over it. If you look straight east from the promontory of Middle Head, the Atlantic Ocean stretches out before you. To the north-northeast lies the Cabot Strait, which separates Cape Breton Island and Nova Scotia from the island of Newfoundland. It is one of North America's busier ferry lanes. Today the enormous car-and-truck-carrying ships of Marine Atlantic ply the strait between North Sydney, Nova Scotia, and Porte-aux-Basques, Newfoundland, year-round through fog and ice and east–west traffic crossing their path, and they have effectively ended Newfoundland's physical isolation from the rest of Canada.

The name Port-aux-Basques first appeared on a map drawn by Samuel de Champlain in 1612, and it acknowledges a Basque presence that probably goes back much earlier than that. The Basques called this harbor Sascot Portu, and it was only one of many that they made their own all up and down the rugged western coast of Newfoundland: Codroy Island between Capes Ray and Anguille, Flat Island in St. George's Bay (which offered good shelter but was also infested with swarms of the blackflies that curse the Canadian North and so was also called by the Basques Ulicillo, the fly hole), Red Island near the western end of the Port-au-Port Peninsula and called by the Basques Isla de San Jorge, Governors Island in the Bay of Islands, Benie Island on the Port-aux-Choix Peninsula, St. John Island, and old Ferrole island, origi-

To Jacques Cartier, the broad course of the St. Lawrence seemed the certain passage through Canada to China. For him and his successors, however, Canada proved just too big an obstacle to go around. (John de Visser)

nally called Ferrolgo Amuirco Punta. There is a fine lighthouse at Ferrole Point, and in clear weather Labrador is visible across the Strait of Belle Isle, and the two hills that Captain James Cook in the 1760s memorably dubbed "Our Ladies' Bubbies."

As early as the 1540s, the Basques hunted baleen whales in these waters, where the plankton-rich Labrador Current follows the north shore of the Gulf of St. Lawrence. During the last half of the sixteenth century, Red Bay in Labrador was the center of the New World's first whale oil boom and saw each year fifteen and more ships from the Basque ports of Spain and France set up shore stations where the great carcasses were brought for flensing and where the blubber was rendered into valuable fuel. For historical archaeologists, the site has yielded a trove of artifacts documenting the Basque enterprise: cooperages, ceramics and glassware, iron harpoons, a cemetery of sixty graves containing 140 skeletons of adult males and two children. Most important were the underwater remains of a whaling ship, probably the *San Juan* sunk late in 1565, and the best-preserved example anywhere of the type of vessel generally used by Europeans to explore the New World.

In the summertime, when the strait is free from ice, a ferry connects St. Barbe in Newfoundland with Blanc Sablon on the Labrador coast and so makes it possible, if one were inclined and had the time, to cross again to the mainland and circle down the valley of the St. Lawrence, through Quebec, New Brunswick, Nova Scotia, onto Cape Breton Island to Keltic Lodge, Middle Head and the twin bays at Ingonish where the Portuguese fished in the 1520s. I say "had the time" because such a circuit would be a daunting trek, approximately fifteen hundred miles. No bridge crosses the St. Lawrence River until Quebec City (seven hundred miles from the Strait of Belle Isle), and on both sides of the great waterway endless miles of rocks, lakes and trees stretch unchanged from the days four and a half centuries ago when the Basques and Bretons first beheld them. Such a journey would convey graphically something of the vastness of the place whose mere edges the fishermen had frequented. The size of the St. Lawrence itself, which dwarfs most of the world's rivers, suggests the immensity of the land it drains. Now, we know just how vast that land—Canada—actually is. Europeans then, until they passed beyond the Newfoundland barrier, had no idea. To know and to master such a place would take more than the practical skills of fishers for whales and cod.

"Pilgrims still go there to pray, make vows, and be shrived; tourists come to admire the site and the architecture—and also consume omelettes chez Madame Poulard ainée. Madame and her rival daughter-in-law are long since gone; but (judging from what their successors did for us) at any hour a fire of crackling thorns will be kindled on an open hearth, eggs broken into a heavy iron skillet, and a delicious omelette produced." That charming image is found in Samuel Eliot Morison's magisterial *European Discovery of America: The Northern Voyages*. It has to do with hospitality

Cartier, like Columbus, died in the belief that what he was really after was China and Japan and not some troublesome land in between, distinguished by illimitable woods and unsurvivable winters. (Thomas Fisher Rare Book Library, University of Toronto)

at the abbey of Mont-Saint-Michel, hospitality that Admiral Morison liked to imagine might have attended the pilgrimage there of King Francis I of France with his son the Dauphin in 1532. Omelettes are a tradition at Mont-Saint-Michel and perhaps one was served to the king and his heir.

As Morison characterizes the meeting, omelettes or no, Francis's visit marked the moment when the history of France in the New World can be said to have begun. The king and his son were received by the abbot, Jean Le Veneur de Tilliers, who in turn introduced to them, with a recommendation, a relative of the abbey's treasurer. The relative was Jacques Cartier, a master mariner from Saint-Malo, born in 1491, and already a veteran of voyages to Newfoundland and Brazil and the man, said the abbot, to lead further expeditions across the Atlantic—if Francis I should happen to have such in mind. He did. This was the monarch who eight years before, in an effort not to let his Spanish rivals get away with all the riches of the New World, had commissioned the Italian Giovanni da Verrazzano to explore the long coastline of America between Newfoundland and Spanish Florida, in search of the still much-believed-in strait through the American barrier and on to the riches of the Indies.

Verrazzano had chosen a route parallel to but 4 degrees north of Columbus's path, setting off from the Portuguese Madeiras in January 1524 and making his American landfall on March 1 at Cape Fear, on the coast of present-day North Carolina. Over the next four months he worked his way steadily up the coast. He mistook Pamlico Sound for the Pacific Ocean, and he missed entirely the opening to Chesapeake Bay. He did come upon New York Bay (but he did not explore it), Narragansett Bay, Casco Bay and Monhegan, Isle au Haut and Mount Desert islands in Maine, was blown past the Bay of Fundy and most of Nova Scotia and on to Newfoundland. He was back in France on July 8, 1524, not having found what he had set out to look for. But he did return with a conclusion that was just as important: that the coast he

The most obvious of the New World's resources, to the eyes of the early discoverers, came from the sea: codfish, seal and whale. The sixteenth century's greatest whale factory was located at Red Bay in Labrador. (Courtesy of the Trustees of the British Library)

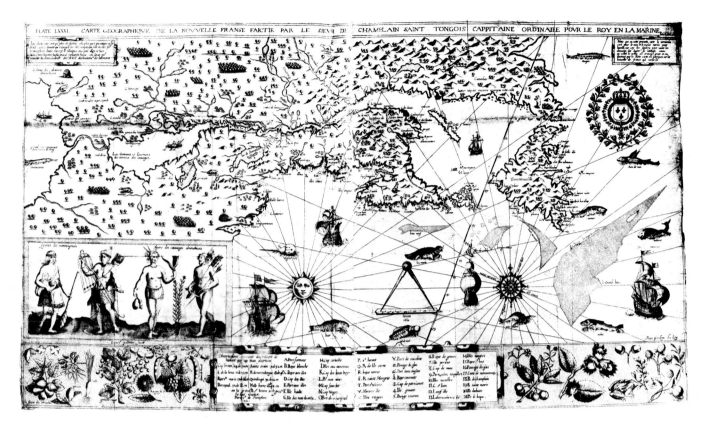

had seen belonged to one single continent and was neither an island nor a promontory of Asia. (Verrazzano hoped to make a follow-up voyage right away, but Francis I just then had European distractions. When he did sail for America again, in 1528, it was with commercial backing and an unhappy outcome: he was killed and eaten by cannibals in the Bahamas.)

Eventually Francis's interest revived, and by the 1530s he was prepared to try again. One result of Verrazzano's reconnaissance was to push later voyages northward to unexplored regions where the elusive strait might yet be found, and so Jacques Cartier, who set sail under Francis's banner on April 20, 1534, headed directly for Canada, or whatever it was that lay behind Terre-Neuve, Newfoundland. Cartier was a favorite of Morison's, himself an avid blue-water sailor who puts the Breton captain in the top rank of navigators and ship handlers of the age of discovery. Cartier made three Canadian voyages and never lost a ship or a man while at sea (though many died ashore), and he explored more than fifty North American harbors without a serious accident. No copy of his royal commission survives, but the narratives of his first voyage make it clear that his charge was twofold: to find a passage to China and to locate precious metals that would make France and Francis rich. The order from the king's treasurer appropriated some six thousand gold francs for equipment and wages for "certain ships which should in company with and under the command of Jacques Cartier make the voyage from this kingdom to Terres Neufves to discover certain isles and countries where there is said to be found a vast quantity of gold and other rich things." Other rich things. Anything, it seems, would do.

Jacques Cartier died at home in France; Samuel de Champlain died in his new home, among Frenchmen, in Canada. By the time Champlain drew his map of New France, the age of discovery had ended and the age of settlement had begun. (Thomas Fisher Rare Book Library, University of Toronto)

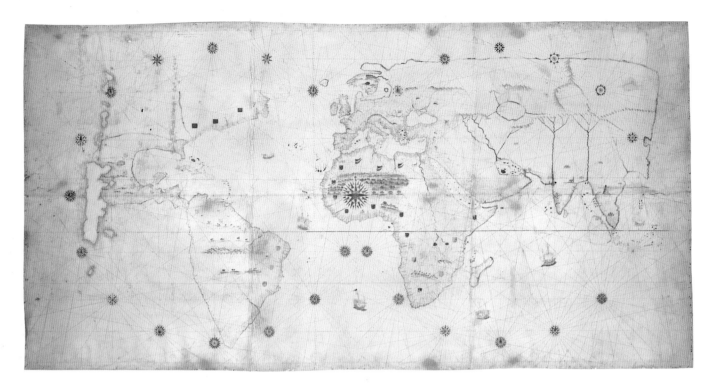

This detail from a 1529 map depicting Verrazzano's voyage was drawn by the explorer's brother, Girolamo, himself a navigator as well as a cartographer. The Sea of Verrazzano is represented but not named, and most of the geographic locations are latitudinally too high. (Biblioteca Apostolica Vaticana)

We know that Cartier had been to Newfoundland before, and that this time his purpose was to sail through the Strait of Belle Isle and into the great gulf beyond. Luck and good sailing gave him a fast passage on the northerly route, and just twenty days out he sighted Cape Bonavista in northeastern Newfoundland. It was still early in the year, and ice drove him to seek refuge, probably in Catalina Harbor, until Westerlies blew away the icebergs and cleared a path to the northwest. The company stopped to kill great auk—the original penguins, now extinct—on Funk Island, where they also encountered and killed a great swimming bear: "Its flesh . . . as good and delicate to eat as that of a two-year-old deer." They rounded Cape Bauld at the top of the Great Northern Peninsula and entered the Strait of Belle Isle. Cartier had some familiarity with these waters from previous voyages but proceeded with caution, following the Labrador coast past Blanc Sablon (where the ferry now lands), Havre de Bonne Espérance, Bois des Hommards, as far as Cumberland Harbor, where he met a lost fisherman whom he put back on course.

He then altered his own course to the southwest (we do not know why) and sailed down the western, back side of Newfoundland, stopping at St. Paul's Inlet on June 16, Cow Head on the Bay of Islands on June 18, and finally sighting Cape St. George and Cape Anguille. Off his port side to the southeast lay the southern passage (the Cabot Strait) back to the Atlantic, whose existence he surmised but did not now confirm. His *Première Relation* noted: "I rather think, from what I have observed, that there exists a passage between Newfoundland and the Land of the Bretons; if so it would be a great time and distance saver if this voyage to [China] succeeds." Instead, he continued westward, exploring the Gulf of St. Lawrence, looking for China. He skirted the

Magdalen Islands and anchored overnight at Ile de Bryon, which though just one mile by five, he described in the hyperbolic tradition of Europeans ever on the lookout for a terrestrial paradise across the seas (or it may just have been that Cartier didn't care much for Newfoundland): "This island was the best we have seen; one *arpent* of it is worth more than the whole of Newfoundland. We found it full of beautiful trees, meadows, fields of wild wheat and pease in flower as fair and abundant as I ever saw in Brittany, and appearing to have been sowed by farmers. There are plenty of gooseberries, strawberries and roses of Provin, parsley and other good sweet-smelling herbs." He also noted bears and foxes and "several big beasts, big as oxen, with two teeth in their mouth like the elephant and which live in the sea"—the walrus.

He moved on to Prince Edward Island, which also impressed him: "All this land is low and uniform, the fairest one could possibly see, and full of fine trees and meadows . . . It is the most temperate land one could ask for" right down to the woodpigeons and turtledoves. Cartier crossed the northern end of the Northumberland Strait to the mainland of New Brunswick, which he followed north to what appeared to be, at last, the opening to China, but was only Chaleur Bay. He anchored at Port Daniel and explored west to Tracadigash Point, near the bay's dead end. Contrary winds then held him at the entrance to Gaspé Bay for ten days (July 16–25, 1534), where he met Iroquois Indians come there to fish from Stadaconé (Quebec City). Observing their apparent poverty ("All their possessions, apart from the canoes and fishing nets, were not worth five sous"), he also noted that "they are wonderful

The arrival of Cartier at Stadaconé, the future site of Quebec City, on September 8, 1535, became a central event in the history and cultural memory of French Canada. This version includes all three ships (La Grande Hermine, La Petite Hermine, L'Emerillon) and at least six banners bedecked with the fleurs de lis of the French kings. (L. Amiel, Arrivée de Jacques Cartier à Québec, Musée du Québec, 80.13, Patrick Altman)

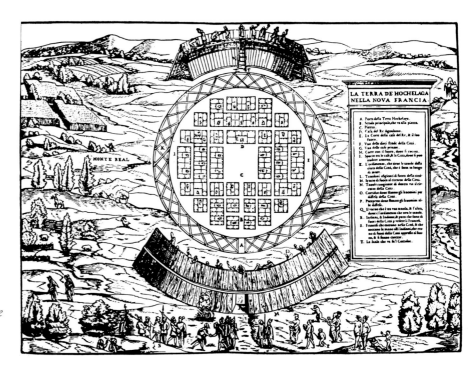

Ramusio's plan of Hochelaga revealed a palisaded Indian city with ten streets, many houses and a special dwelling for the king. (Metropolitan Toronto Reference Library)

thieves, filching anything they can lay hold of." Their chief, Donnacona, presented two of his teen-aged sons, Domagaya and Taignoagy, for Cartier to take with him back to France, and probably gave Cartier the welcome he did more from his own need for an ally against hostile tribes than out of fascination with the French. Then there followed at Gaspé one of the great set-pieces of discovery imagery, as Cartier's men raised a very noticeable token of possession: a thirty-foot-high wooden cross and shield, emblazoned with three fleurs-de-lis and inscribed portentously "Vive Le Roy de France."

It was the high point of the voyage. Dense fog prevented Cartier from rounding the northern shore of the Gaspé Peninsula and moving up the St. Lawrence, and instead he sailed around singularly unpromising-looking Anticosti Island. By then it was August, and choosing not to risk going farther and spending the winter in an unknown land, he turned his two ships (which remain nameless) for home. He passed out into the Atlantic through the Strait of Belle Isle and made Saint-Malo on September 5, his reconnaissance complete.

The first voyage set him up for another, which went farther and attempted more. This second expedition to Canada, in 1535 and 1536, also sailed under the royal warrant of Francis I and was a larger affair: 110 men plus the two Indian boys, in three ships, *La Grande Hermine, L'Emerillon* and *La Petite Hermine* (*Hermine* means weasel, one of which appeared on the arms of Anne of Brittany, queen of Louis XII), provisioned for eighteen months. The westbound voyage was not easy. They sailed on May 19, 1535, but were separated by storms and did not make land for fifty days. On this trip Cartier did not bother with Newfoundland at all but, after gathering two boatloads of great auk on Funk Island, made straight for Blanc Sablon on the Labrador coast and thence directly for the great gulf. On August 10 he took shelter in a small harbor (now Pillage Bay) on the northern shore, which he named La baye sainct Laurins (that date being the Feast of St. Lawrence). It was the first appearance of that name which later Frenchmen would attach to the gulf, the river and the mountain range off to the north. From there, he followed his two Indian guides up the great river of Hochelaga (as they called it) to their home at Stadaconé. En route he confirmed that Anticosti was an island, and he tacked between the river's banks in search of anything that might be "the strait."

The Indians also had told him of the riches of the legendary Kingdom of Saguenay, which they said lay to the north between the Saguenay and the Ottawa rivers and which became the primary object of Cartier's desire during all his remaining time in Canada. At Stadaconé, the two Indians returned to their people, and Cartier sought out a safe long-term anchorage for his two largest ships, since he would have to explore upstream in smaller boats. He found it at the place where the St. Croix (St. Charles) and the Lairet rivers meet, and there he laid up *La Grande* and *La Petite Hermines*. With some fifty of his crew, he set out on September 19 upstream toward the Indian town of Hochelaga. There on October 3 the Iroquois welcomed the French with courtesy in their citadel beneath a great hill. Cartier named it Mont Royal, and the name stuck.

From its crest, where the Indians conducted their visitors, the Frenchmen beheld a view of the Laurentians on the north, the Adirondacks to the south and, to their great distress, up the river a chain of rapids that it was obvious no European boat could pass. They are called the Lachine ("China") Rapids—so named in the next century by René-Robert Cavelier de La Salle (and not by Cartier as the legend has it) as a joke on the fact that this was as close as Cartier ever got to China. The Indians spoke tantalizingly of the river becoming navigable after three more sets of rapids were passed, of the great wealth in the Laurentians along the Ottawa River and, back the other way, of the riches of Saguenay. Cartier diligently noted it all down and then returned to Stadaconé.

He arrived on October 11 and settled in for what turned into a very long winter. We do not know if it was visions of Saguenay that sustained him, but something quite extraordinary must have. For a ferocious Canadian winter soon struck the unprepared Europeans full bore and must have made him wonder how it was possible to live in such a place. "From the middle of November until the fifteenth of April, we lay frozen up in the ice, which was more than two fathoms in thickness, while on the shore there were more than four feet of snow, so that it was higher than the bulwarks of our ships," Cartier wrote. "All our beverages froze in their casks. And all about the decks of the ships, below hatches and above, there was ice to the depth of four finger-breadths." By the end of the year scurvy had broken out and ravaged almost everyone except, miraculously, Cartier himself. Only an Indian potion made from a tree called *annedda*—the common arborvitae—saved them, and 85 of the company of 110 managed to survive till spring.

Scriue anchora il medesimo messer Guielmo, che se ne trouano nelle fattezze del corpo alquanto differenti dal predetto : niente di manco sono vecchi Marini, & nell'Indie Occidentali chiamansi lupi Marini. ha il corpo con tutte l'altre parti piu grosso, & in se piu raccolto, che non ha il sopradetto.
Vecchio Marino nell'Oceano, Lupo Marino nell'Indie,

Cartier's Première Relation *offered detailed observations on his first voyage of 1534, including descriptions of odd new sea life like seals and otters. (Metropolitan Toronto Reference Library)*

In July 1534, Cartier explored Chaleur Bay and made contact with the Micmac Indians. The Micmacs later became accustomed to Europeans, and traded furs for items such as firearms, as shown in this anonymous 1850 painting. (National Gallery of Canada, Ottawa)

Donnacona meanwhile, obviously a yes-man and yarn-spinner, continued to feed the ever eager-to-listen Frenchmen with ever more fantastic tales about the immense quantities of gold, rubies and other precious commodities of Saguenay, which, he said, could be approached up either the Ottawa or the Saguenay rivers. And he claimed himself to have been there.

That perhaps was unwise, for just before the French pulled out in May 1536, Cartier laid a successful plan to kidnap the chief to witness to his royal patron, Francis, of the Mexico-like wealth that just waited unlocking in Canada. Cartier also seized Donnacona's two sons and two important companions, who joined the several children the Indians had already offered as gifts. He promised to return them all in "ten or twelve moons," laden with presents from the French king; none ever saw Canada again. As if in a peace offering, however, Cartier left behind the hull of *La Petite Hermine* (which he no longer had the crew to sail), rich with the iron nails that the Indians treasured more than gold. And with that gesture, Cartier and the French were gone for a second

Cartier arrived at the site of Montreal, or Hochelaga, as the natives called it, on October 2, 1535. A detail from this highly romanticized nineteenth-century rendering of the event partakes of timeless discovery cliches: white men bearing brightly colored gifts; naked red men and women bearing thick furs; an impressively conjured natural setting, in this case, Mount Royal itself. (Eugène Hamel, L'Arrivée de Jacques Cartier à Québec, Musée du Québec 34.233, Patrick Altman)

time. They set sail on May 6, passed between Anticosti and the Gaspé Peninsula on May 21, sighted Cartier's favorite Ile de Bryon three days later, and then made for home through the passage between Newfoundland and Cape Breton Island that he had guessed at on the first voyage. On Whitsunday, June 4, they put in briefly at a Newfoundland harbor just east of Cape Ray (probably Port-aux-Basques) and on June 11 called at Les Iles de Saint-Pierre, where they met Breton fishermen, the first ships they had encountered since leaving France fourteen months before. They reached Saint-Malo in mid-July, with the Indians and a mother lode of notes about the resources Cartier believed Canada held out to Frenchmen willing to go after them. He also returned with detailed, if incomplete, knowledge of the water route into the North American interior that in the next century would become the lifeline of New France.

Cartier's third and final voyage did not take place until 1541, and by this time hope of getting through Canada to China had just about vanished. But the prospect of Canada's mineral riches beckoned more strongly than ever. Donnacona (who was presented at court, baptized, given a pension and eventually a Christian burial), it seems, did a good sales job on the king. Preparations were elaborate and took three years, but in an odd twist Francis placed in command, over Cartier, Jean-François de la Rocque, Sieur de

Known as the Master Mariner of St. Malo in Brittany, Cartier left no contemporary portrait, but he earned his reputation anew on his three voyages to Canada, where he explored some fifty harbors and did not lose a single sailor to the hazards of the sea. This painting is a modern interpretation of Cartier's departure from St. Malo. (National Archives of Canada)

Roberval, a high-born nobleman and a Protestant. Rumors that the French at last were mounting a serious colonizing expedition to the New World set European diplomatic circles buzzing. Charles V of Spain even ordered special fortifications built at Havana, San Juan and Santo Domingo to repel expected attacks by the French fleet.

But no harm was done: the French aimed only to return to Canada. Though Roberval was technically in command, he had never been to sea, and it was agreed that Cartier, who was ready to sail first, should proceed ahead with part of the expedition. He left Saint-Malo on May 23, 1541, with five ships, men, women and livestock. Three months later they were at Stadaconé, where Cartier reported that the Indians gave them a fine welcome—even though Cartier had conspicuously not made good on his promise of 1536 to return in a year with Donnacona alive, well and bedecked with French honors. This time Cartier made his base at Cap Rouge, eight miles upstream from the Indians' capital, perhaps sensing that beneath the surface the natives were not especially glad to see him again. There he put his colonists and convicts (he had had to sweep the jails to fill out his complement) to work on a palisaded settlement, which he named Charlesbourg-Royal after the king's son, Charles, Duke of Orleans. Others he set to collecting samples of local gold and diamonds to send back to France on two of the ships that were to return on September 2. Cartier and his own select party of adventurers wasted no time in setting out for Hochelaga, which they reached in small boats on September 11 and where they obtained Indian guides who could supposedly take them to Saguenay. It was a short-lived expedition. The Indians led Cartier only as far as a small village on the north bank of the Lachine Rapids and indicated that yet another such *sault* still lay ahead. There the record stops; Cartier went no farther.

Back at Charlesbourg-Royal, the French prepared for another winter in Canada, but this time not just the climate but the natives too threatened to undo them. Of this last winter that Cartier spent at the site of Quebec City there is no extant account. Talk of those who survived indicated that thirty-five French were lost to the Indians, that scurvy again broke out and was again cured by the magical arborvitae, that the cold was intense. Safe to say, there were no omelettes to eat, and in 1542 when the ice went out the French did too, but not before loading onto *La Grande Hermine*, *L'Emerillon* and the third ship eighteen barrels of gold and silver (iron pyrites) and a basket of rubies and diamonds (quartz crystals). On the way home, Cartier called at St. John's in Newfoundland, where to his chagrin he found Roberval's much-delayed part of the expedition heading the other way. His superior instructed Cartier to turn around and accompany him back up the St. Lawrence, but Cartier, who had had enough, "stole privily away" and returned to Saint-Malo after an absence of sixteen months.

Roberval forged ahead alone, but despite being well armed and provisioned, he spent just one discouraging scurvy-ridden winter among the sullen natives at Cap Rouge (which he renamed France Roy) before he called it quits. He too made a quick stab at Saguenay and Lachine Rapids but got nowhere either—because there was nowhere to get to, no Mexico of the North, only tales told by guileful savages to gullible white men. "Saguenay" lingered as a place reference on the maps for another hundred years; today it survives only as the name of the river that flows into the St. Lawrence at Tadoussac.

The Lachine Rapids on the St. Lawrence above Montreal, as shown in a nineteenth-century drawing, marked the limit of Cartier's penetration of the Canadian interior. The name ("China Rapids") was a subsequent joke on the fact that this was as close as the Frenchman from St. Malo ever came to the Orient. (Metropolitan Toronto Reference Library)

Prior to the arrival of Europeans, the site of Quebec City was occupied by the Iroquoian village of Stadaconé. Cartier probed the reaches of the St. Lawrence, hoping to find access to China. (Guido Alberto Rossi/The Image Bank)

The reasons for Roberval's decision to abandon Canada seem obvious enough. The job must have looked to be enormous, and he must have judged that, back home, the fire-in-the-belly desire for a North American empire needed to sustain such an enterprise over such vast distances simply was not yet there. No record remains of his thoughts, only a prosaic inscription on a map of the region: "It was impossible to trade with the people of that country because of their austerity, the intemperate climate of said country, and the slight profit."

His intuition proved correct, for in France things changed in ways that were not auspicious for overseas enterprise. Francis I died in 1547, and under his son, Henri II, the kingdom fell into bloody religious wars that sapped her energy for decades. Roberval died in a religious riot in Paris in 1561. Cartier, who had survived scurvy in the New World, died in his Saint-Malo manor house in 1557, probably of the plague. And Canada would not be bothered by Frenchmen again until the age of discovery had become the age of settlement. When Canada was intruded upon once again, the intrusion would be permanent and uninterrupted, and it would be accomplished not by a great sailor like Jacques Cartier but by a great empire builder like Samuel de Champlain, whose country this time was sure of what it wanted.

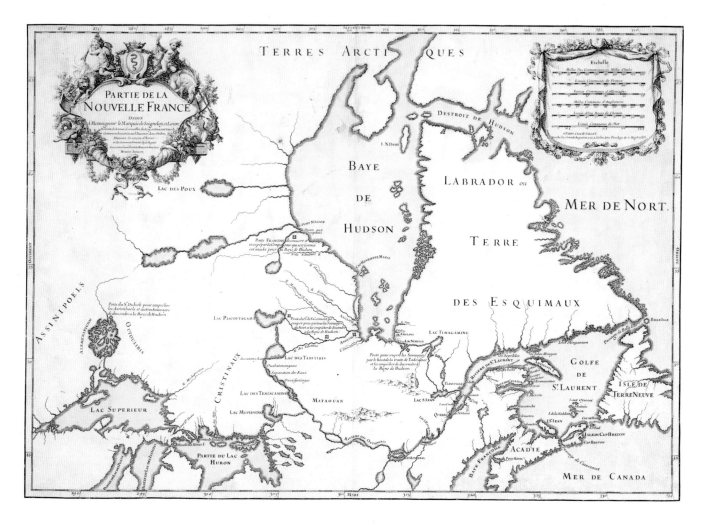

Jacques Cartier made three voyages to Canada and died in Brittany. Samuel de Champlain crossed the Atlantic twenty-three times between France and Canada and died in Quebec. His story, which is the story of the successful colonization of the land—Canada—that Cartier had discovered, is beyond the limits of this book except in this regard. As the "Father of New France," Champlain consummated what Cartier had begun. Like Cartier, Champlain sailed in the name of the king of France, and after him there would always be Frenchmen in Canada.

In Montreal (née Hochelaga), the Canadian Pacific Railway in the 1960s built a sleek and beautiful hotel and called it Le Château Champlain. It joined a long line of great hostelries that the CPR had put up along its transcontinental mainline since its completion in the 1880s: the Empress in Victoria, the Banff Springs in Banff, the Saskatchewan in Regina, the Royal York in Toronto, the Château Frontenac in Quebec. In the years after confederation in 1867, which established Canada as a self-governing dominion within the British Empire, the Canadian Pacific Railway at great cost had been punched across the greatest physical obstacles to true national union: the Canadian Shield, the Great Plains, the Rocky Mountains. When the last spike was driven home, with it went the hopeful message that Canada, *a mari usque ad mare*, was one country. Ever since, the transcontinental railway has had

New France as it developed in the seventeenth century embraced territory of vast dimensions and riches. Though France could not ultimately hold onto her empire in Canada, the culture she planted there remains resiliently French to this day. (National Archives of Canada)

as much symbolic as practical value, and the great hotels all along its length have punctuated that brave declaration of Canadian nationhood.

Le Château Champlain was a latecomer in all this, but it is still faithful in one detail at least to that palpable sense of history that surrounds its dowager sisters. In the lobby hangs a grand portrait of its namesake. By Canadian artist Marc-Aurèle De Foy Suzor-Coté (1869–1937), it is large and colorful and fills a wall near the concierge's desk. Few guests pay it much notice, and the brass plate says it was presented to the hotel by T. R. McLagen, chairman of the Canada Steamship Company. Though I do not know its history, it is the commanding sort of picture well suited to a company boardroom—or to the grand saloon of the Canadian Pacific's steamships (ships as well as trains and hotels went into "The World's Greatest Travel System"): *Empress of Canada* or *Empress of Britain* or *Empress of England* or *Empress of Scotland*, which into the 1960s showed Canada's flag (not the modern maple leaf but the old one with the Union Jack) on the North Atlantic sea-lanes to Liverpool. The picture is a nod to history in an otherwise very modern setting; Montreal is a modern city, modern enough to have self-consciously preserved part of its riverfront and called it Old Montreal. It is also French, and faithful in language at least to the memory of Champlain, whose name graces a large bridge across the St. Lawrence.

The best view of Montreal is still to be had from atop Mount Royal, where, just a century ago Francis Parkman, the greatest American historian of the nineteenth century and one of the greatest of any age, wrote this about it: "From the summit, that noble prospect met his [Cartier's] eye which at this day [c.1885] is the delight of tourists, but strangely changed, since, first of white men, the Breton voyager gazed upon it. Tower and dome and spire, congregated roofs, white sail and gliding steamer, animate its vast expanse with varied life. Cartier saw a different scene. East, west, and south, the mantling forest was over all, and the broad blue ribbon of the great river glistened amid a realm of verdure. Beyond, to the bounds of Mexico, stretched a leafy desert, and the vast hive of industry, the mighty battleground of later centuries, lay sunk in savage torpor, wrapped in illimitable woods." Parkman's "modern" Montreal of the 1880s is what is now preserved in Old Montreal. His picture of what the country was like before the French came and crafted Canada out of its illimitable woods captures the frightful immensity of that job. For Cartier it all proved too much; for Champlain it proved just possible; for their Canadian successors the job of getting Canada together, and then keeping it together, has proven no less immense.

That challenge is due to enduring obstacles to national unity that are not of the geographical kind once overcome by the building of a transcontinental railway. Today the obstacles are cultural, and partly political, and only a collective loss of historical memory could erase them. At Quebec City (née Stadaconé), where Cartier tried in vain to make a go of it in Canada, the signs are that that memory is far from dead. Like every other city today, Quebec can be reached by air, but it is not the best way to come. The train from Montreal will carry you gloriously 150 feet above the St. Lawrence on the 3238-foot Quebec Bridge, opened in 1917 and the longest cantilever span in the world. Or you can arrive as Cartier did, on the water. There are no great steamships anymore to bring you here from the old countries, but there is a ferry from the town of Lévis on the south bank of the river, and it makes for one of the most romantic urban approaches anywhere in the world.

The scale of Quebec is still reasonably modest, and while its skyline is etched with the standard modern and postmodern pillars, it retains the character of the fortress that it originally was. The great Rock of Quebec rises between the St. Lawrence and its tributary St. Charles, and from Cartier's time onward has resounded with names that made history and were not forgotten: Champlain, Bishop Laval, Count Frontenac, Generals Montcalm, Wolfe and Montgomery, Canadian premiers Wilfrid Laurier and Mackenzie King, United States president Franklin D. Roosevelt, British prime minister Winston Churchill. But mostly, Quebec remembers the Seven Years War (the French and Indian War, as it was called in America) and the early morning of September 13, 1759, when British soldiers under General James Wolfe made their surprise ascent to the Plains of Abraham above the walled city and defeated the French under Louis Joseph, Marquis de Montcalm. Wolfe and Montcalm died in the encounter, Quebec soon capitulated, and within a year all Canada was British. The moment was made famous for Englishmen by Benjamin West's great painting, "The Death of Wolfe." For the French it needed no dramatization: with it ended 151 years of French rule in Canada, if you count from July 1608 when Champlain raised the fleur-de-lis at Quebec.

The walled city and the fortress are still there. Since 1920, the Citadel (an elaborate fortification 350 feet above the river built by the British in 1820 on old French foundations) has been in the care of the Royal 22nd Regiment of the Canadian Army. More affectionately known as the Van Doos (as in *vingt-deuxième*), it is Canada's only all-French-speaking regiment and is the source of considerable local pride. No free wandering about the Citadel is permitted, but serious university student guides will show you around the battlements and into a few of the old stone buildings. Some tours are in English; some in French. Everything is very spit-and-polish; the busby-hatted guard changes snappily; the mascot Himalayan goat, Batisse V, stands to with gold-painted hoofs and horns, his coat and teeth brushed daily.

Yet a strange atmosphere hangs over the Citadel today that is apparent in what is said and how it seems to be heard. It derives from the history that began here when Cartier discovered the great promontory in 1535 and that ended decisively in 1759 when French power again proved insufficient to hold on to it. The motto of the Van Doos, which appears in various forms all over the Citadel

Samuel de Champlain, known as the Father of New France, consolidated and rendered permanent in the seventeenth century the Canadian discoveries made by Cartier in the sixteenth. (National Archives of Canada)

and on the uniforms of the soldiers themselves is *Je me souviens*, literally "I remember," or more poetically—"Lest we forget." When I took the tour an elderly anglophone gent, who from the look of nostalgia in his eye may have been a World War II veteran, asked what that meant. He probably had in mind that it must refer to the memory of some past battlefield glory, some Belleau Wood or Omaha Beach. "Nothing so small," the young francophone guide quickly set him straight.

If the motto must be said to refer to a single event, that event is Wolfe's defeat of Montcalm on the Plains of Abraham just over the wall, which many French Canadians still interpret as one of history's monumental injustices, ever to be remembered as "the Conquest." We must never forget that we are French, said the young woman guide, and she said it not as a guide routinely reels off details about the cannons, the bayonets and the powder magazine, but as a believer with passion, pride, perhaps a trace of contempt. I have never seen a place that in appearance seems so utterly British (the whole setting recalls Edinburgh Castle, and it is nicknamed the Gibraltar of America) but in feeling seems so un-British. Though the Van Doos was designated a royal regiment by George V and though the Quebec residence of the governor general (Her Majesty's representative in Canada) sits squarely inside the fortress, the Citadel is no "little bit of England." It is pure Canada, and betrays the unease of that vast country, divided by language, a large part of it nursing its ancient grievance.

Best to move on. Turn right off Côte de la Citadelle at Port Saint-Louis and walk down rue Saint-Louis to the other great Quebec monument, which like the fortress is also a working site: Canadian Pacific's castellated Château Frontenac, named after Louis de Buade, Comte de Frontenac, governor of New France from 1672 to 1682. An imitation château whose grand scale rivals the originals back in France, the place is usually abustle with Québécois and with visitors drawn to the charms of Old Quebec, since 1985 a UNESCO World Heritage treasure. I do not know if the UNESCO men when they came to town stayed here, or if as they passed through the baronial lobby they looked up. If they did not, then they missed a real treasure: six leaded and stained-glass windows overlooking the entrance courtyard beyond. Each memorializes a chapter in the history of discovery: "George Vancouver Discovers Vancouver Island in 1792"; "The Brig *Hector* arrives at Pictou, 1773"; "Zachariah Gilliam reaches Hudson Bay [date illegible]"; "Jacques Cartier Sails up the St. Lawrence, 1535"; "Columbus Sets Sail for America, 1492." (Note the subtle Canadian touch: Columbus only set sail, but Cartier discovered.)

It is not unusual for buildings conceived and constructed with the care that this one obviously was to have hidden about them in surprising nooks and crannies such wonderful details as these. In their ability to give the traveler pleasure, I venture that they greatly excel the mere portrait on the wall, the ostentatiously planted suit of armor, the always-too-large coat of arms. As evocations

A Van Doos guard, from the Canadian Army's Royal 22nd Regiment, stands at attention in the Citadel. (John de Visser)

of the age of discovery in Canada, they certainly excel the official presentation of the subject, out at the Cartier-Brébeuf National Historic Park along the St. Charles River, where the French passed their first horrific Canadian winter.

This is nevertheless the actual site where Cartier moored his ships, made his notes and watched the snow pile up and his men drop from scurvy. There are helpful exhibits in the visitor center and bilingual Parks Canada guides to show you about.

In addition to Cartier, you learn about Jean de Brébeuf and the Jesuit fathers who came to establish the Mission of Notre-Dame-des-Anges and carry the faith to the Huron. And, out in the park, there is the ship: against the skyline of Quebec, itself a gay mix of château cupolas and office towers, rise the three masts of Cartier's "Big Weasel," *La Grande Hermine*. The national park service commissioned construction of the replica for the centenary of Canadian confederation in 1967 and spared nothing to see that every plank and nail were as close to the originals as you could get four hundred years after the first *Grande Hermine* ended her days with Cartier back in Brittany. The École Polytechnique de Montréal did two years of research in Canadian and European libraries; naval architect François Cordeau directed construction at the Davie Brothers Yard at Lauzon, Quebec; craftsmen labored long using traditional techniques to produce an authentic mid-sixteenth-century French ship.

But the place does not seem to be a big tourist draw. It was not high on the list of cultural attractions recommended to me by the concierge at the Château Frontenac, and, sitting out on the St. Charles, it is a good taxi fare from the tourists' ground-zero in Old Quebec. The cabbie, when told, in French, to go to Cartier-Brébeuf Park, looked puzzled, but then turned and asked, in English, "The old sheep?" And we were off.

It is a very fine new old ship, and I and just three others had it to ourselves. No rush, lots of time for questions, plenty of opportunity to pass your hands along the ribs and planking, to bump your head on the beams. The speech, as seemed fitting in modern Canada, was trilingual: English for me, French for one of the others, who then put it promptly into Portuguese for his two guests. I have no Portuguese and so cannot say whether what his guests offered in return made any reference to Portuguese pre-eminence on the coast of Canada before the French ever got serious about it.

There is always something disconcerting about a ship out of water. Although the dated pictures and guidebook descriptions would lead you to think *La Grande Hermine* rides placidly at anchor in the St. Charles, she is, sadly, high and dry, scaffolded and up on stilts, her twenty-two-year-old timbers rotted, her upper deck closed off. She seems to have aged a little faster than *Susan Constant*, *Godspeed* and *Discovery* down at Jamestown, which are ten years older and still afloat, but which also must soon give way to a new set of replicas.

Perhaps it is those Canadian winters.

The replica of Cartier's flagship La Grande Hermine *was lovingly constructed for the occasion of the hundredth anniversary of Canadian Confederation in 1967, and sits today in a Quebec City Park, propped up out of the water, her timbers badly rotted. (John de Visser)*

Parris Island

ACROSS THE SALT MARSHES OF LOW COUNTRY South Carolina, from Beaufort and Port Royal, Highway 281 leads to Parris Island. Traffic is heavy, for nearby the resort of Hilton Head, whose highly refined and highly profitable business is leisure, beckons all manner of the old monied and the upwardly mobile. But not Parris Island, whose product, as the slogan on the post office bulletin boards puts it, is "a few good men" (and today some women too), and which beckons only to a few of America's youth, rich and poor alike. USMCRD Parris Island: United States Marine Corps Recruiting Depot (Eastern Region), the place they come to if they live east of the Mississippi River and think they've got what it takes to become a leatherneck.

At the outer gate, a starched sentry waves you onto the base, never once breaking the practiced rhythm of the corps salute. An orderly parade of vehicles follows and precedes you: civilian workers heading for jobs on base (groundskeepers, check-out clerks at the commissary, chefs at the base Burger King), marines returning from off-base recreation, gawkers like me. Cars move swiftly past the sentry box and onto the long two-lane road built up over the marsh, a man-made isthmus that makes the "island" into an accessible peninsula.

Once on the island, the drive winds under live oaks and then debouches into the functional, government-issue landscape of an obviously military establishment. However a Marine ends up—happily retired at Hilton Head after twenty years' service or blown up in Beirut not yet twenty-one—Parris Island is for him where it all begins. He can't miss it; nor can you. Just read the enormous welcoming sign that bridges Boulevard de France near the parade ground where recruits so proudly march for their families on graduation day: "USMCRD Parris Island—Where It All Begins."

But I am not a Marine, nor am I related to one, and what took

Elizabeth I's most famous captain, Sir Francis Drake, plundered Spanish St. Augustine in Florida in 1587 (though he spared its inhabitants), and frightened off the Spanish from Santa Elena for the last time. Thenceforth, Carolina's colonial history would be English. (Metropolitan Toronto Reference Library)

The flora of the New World amazed and delighted Europeans long used to their own highly domesticated landscape. To them, the new American wilderness could be as lovely — in the bloom of the dogwood, laurel, jessamine and wild rose — as it was sometimes frightful. These early paintings are by the English artist, Mark Catesby. (Courtesy of the Trustees of the British Library)

me to Parris Island was the search for other beginnings. Centuries before this South Carolina sea island became the place "where it all begins" for Marine recruits, it was the place where it all began for other sorts of recruits on missions with odds as long as any the Marines today are taught to surmount.

Just follow the driving tour out past old Page Field where Marine Corps pilots once took off to hunt German submarines along the Carolina coast and where Navy dirigibles once docked, past the Ammo Supply Dump (Restricted To Military Personnel Only), to the carefully manicured greens and long fairways of the base's eighteen-hole golf course. Turn left at the clubhouse parking lot and go a short distance to a small circle from which three historical markers are visible.

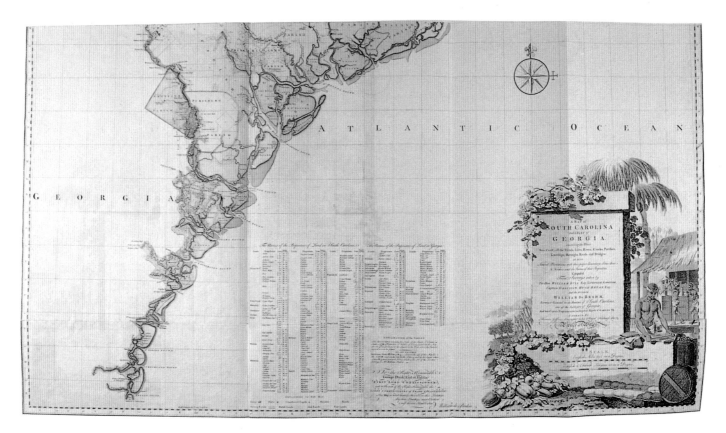

What they record and what is elaborated on in the base museum in the War Memorial Building is a particular instance of the larger process of discovery: a tale of flight from oppression, of greed for gold, of rivalry between jealous earthly powers and exclusive religious creeds, of the deadly meeting of Renaissance Europeans with Stone Age Americans. As was surely not the intention of the faithful who erected them, the three markers commemorate the confusion that attended Europeans' efforts to do something with the new land on the western side of the Atlantic Ocean where they had hoped, and many still believed, China should rightly be.

By the time anything worth noting ever happened on Parris Island, the existence of the great American landmass had been known to Europeans for sixty years. But their epic voyages of discovery, which had begun in the 1490s and which by the 1580s had pretty well etched the coastline of the Americas from Baffin Island to the Straits of Magellan, waited (in North America at least) for other forces to render those discoveries permanent. At that point—the point at which places like Parris Island enter the historical record—the age of seaborne discovery verges into the subsequent age of land-based settlement.

Bear in mind that even by the late sixteenth and early seventeenth centuries, when the first permanent settlement efforts took hold, the ability of the fledgling nation-states of late Renaissance Europe to carry out and sustain colonizing enterprises at such vast distances was still marginal at best. The vagaries of nature and human behavior and the limits of the technology of the time dictated a certain randomness to these efforts and should remind us of the unlikely and utterly precarious nature of their history.

We should also be reminded that the story of discovery, at its

The civilization that had its earliest troubled beginnings at Charlesfort and Santa Elena in the sixteenth century, bloomed by the eighteenth century into the wealthiest and most polished of Britain's American colonies: South Carolina with Charleston at its heart. This map from 1757 details the reach of the Low Country domain. (The Newberry Library, Chicago)

most modern, is not a modern story at all, and that the men who moved it forward did so only hesitantly with feet more firmly planted in the certainties of the Old World than with eyes intent on the horizons of a new one. They did not sow the seeds of our own later democracy; our technocratic egalitarian society would have seemed as remote and undesirable to them as the Stone Age one they confronted. We now look back over nearly half a millennium after the key events of discovery took place, events at the dawn of a modern age that they partly precipitated but whose consequences no one then could or would have wanted to foresee.

There's not a lot to look at out by the golf course at the tip of Parris Island, and it's probably safe to say that few people bother to take in the view. You will find no costumed guides, no reconstructed fort or village, no National Park Service visitor center. Only a rectangular space marked by concrete stakes at the edge of a wood overlooking open marshland and tidal rivers. With thick humidity, temperature in the low nineties, and only the slightest stirring of a breeze, the only company you will find here are little lizards flashing noisily across ground covered with the dried brown leaves of live oaks. Spanish moss hangs spectrally from low branches, and white sand and seashells crunch underfoot.

European feet first trod Parris Island in the early 1520s—St. Helena Day, August 18, 1521—when Spaniards probed the northern edges of the new world that, thirty years before, Columbus had so blazingly brought to the Old World's attention. It marked the beginning of their touch-and-go efforts to establish outposts on the southeastern coast of the North American mainland. They reported the existence of a great harbor, which they called Santa Elena and which the French and English called Port Royal, and of a hospitable landscape. They did not linger, how-

The construction of Charlesfort by French Huguenots somewhere on Parris Island in the 1560s marked the effective beginning of Low Country history. Though at first unsuccessful, the Huguenots later came to play a major part in the development of Carolina culture. (Parris Island Museum, USMC)

ever, and for the next decades Spanish attention remained fixed steadfastly on central and southern America, where the riches they had come for awaited their seizure.

In that respect, northern America would ever prove a distinct disappointment for treasure-hungry Europeans. It is tempting to venture that Spain's delay in further settlement efforts north of Florida (until it was too late and others had a foothold) was the result of an intuition that the northern reaches of whatever world this was would demand, to pay off, harder work and greater patience than the average conquistador routinely packed in his kit. In any case, it took an intruder—the threat of competition—to bring the Spanish back to Parris Island.

The threat came from France and was both political and religious. Huguenots (Protestants) from France's maritime provinces sailed from Le Havre in February 1562 and in May landed in Port Royal harbor. Sponsored by Gaspard de Coligny, Admiral of France, and led by Jean Ribaut with 150 men, the French made first land-fall near the site of the future city of St. Augustine, which Ribaut felt was too close to Spanish territory. Moving north up the coast of what would one day be South Carolina, he renamed Santa Elena Port Royal, and in token of possession erected on Lemon Island a pillar engraved with the royal arms of France.

Partridges, wild turkey and other game abounded; the natives were welcoming. Ribaut asked for volunteers to stay behind while he fetched supplies and reinforcements from France. French motives encompassed greed and altruism: they sought a place for Protestants unmolested by Catholic reaction, and a military staging post for French attacks on Spain's treasure fleets. Frustrations vexed them on both counts.

When Ribaut left the twenty-six men at the little settlement on Parris Island called Charlesfort (after Charles IX of France), trouble soon followed. The settlers mutinied against their leader, built a makeshift boat and attempted to sail home to France. Few survived. Ribaut did make it back home, but found on his return a France newly riven by religious conflict, its monarch distracted and for the time uninterested in a resupply mission to Charles-fort. Desperate, Ribaut decamped in 1563 across the Channel to England, where he published *The Whole and True Discovery of Terra Florida*, and appealed for aid to the Protestant Elizabeth I. But the great queen's enthusiasm for discovery, soon to be so vital to the enterprises of Hawkins, Raleigh, Drake and others of her captains, was still unfocused, and Ribaut went away unsuccored.

Meanwhile a Spanish expedition dispatched for the purpose by Philip II had found out and destroyed the by now uninhabited remnants of Charlesfort. In 1564 Admiral Coligny tried again, sending a veteran of Ribaut's first voyage, René de Laudonnière, back to America, where he established Fort Caroline (again, after Charles IX) on the St. John's River in north Florida. Again dissension, dis-organization and the distractions of treasure-hunting sundered the settlement, and as food supplies dwindled the Frenchmen once more

Northwest Passage

WHILE THE SPANISH AND THE FRENCH WERE TRYING TO establish outposts on the southeastern coast of North America, explorers pressed forward in their eagerness to find a short route to China over the top of the American landmass. In his book published in 1576, *A Discourse to Prove a Passage by the North-west to Cathaia and the East Indies*, Sir Humphrey Gilbert said that America was really the continent of Atlantis and that the Strait of Anian reached around its northern side toward China. Gilbert himself did not try to find it, but two other Englishmen did and returned to tell about it.

In 1576, 1577 and 1578, Martin Frobisher, who was an experienced sailor and sponsored by Gilbert and the Earl of Warwick, made three voyages northwest past Cape Farewell at the southern tip of Greenland to Baffin Island where he explored Frobisher Bay and Hudson Strait. He thought he had found the strait to China, but of course he hadn't. He also thought he had found immense gold deposits and carried back to England two hundred tons of worthless iron pyrites that assayors in London confirmed were quite genuine. On his second voyage he took with him artist John White whose watercolors gave Europeans their first glimpse of the Eskimos, with whom Frobisher had several hostile encounters.

John Davis, who was also backed by Gilbert and William Sanderson, a London merchant, sailed in search of the great passage first in 1585. From the west coast of Greenland, he ventured across the strait that bears his name to Baffin Island. He believed Exeter Sound to be the passage, but was forced to abandon the search by the lateness of the season. He made a second voyage the following spring, skirted but did not enter Cumberland Sound and then followed the coast of Labrador south where he gathered a cargo of cod and sealskins before returning home. On his third and final voyage he noted the mouth of the Hudson Strait. That strait was named for Henry Hudson, the Englishman who

The desire to trade led sixteenth-century Europeans east as well as west, near the top of the world as well as around its middle. Willem Barent's map of 1598 sketched European probes through the polar regions to the top of Russia. (National Archives of Canada)

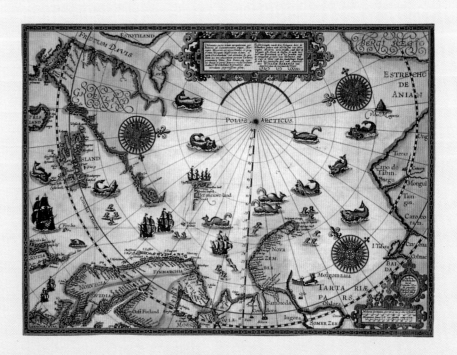

in 1610 penetrated it to the great interior bay that also bears his name, to be marooned there by a mutinous crew in 1611.

Five years later Robert Bylot and William Baffin concluded that the way to China would not be found via Hudson Strait but via Davis Strait, Smith, Jones and Lancaster sounds. They were right, but the way would not be proven for another three centuries, and it never would be of any practical use in getting to China.

opted for home. But before they could get away, Ribaut reappeared with seven ships and more colonists. Sadly, so did a Spanish expeditionary force led by Pedro Menendez de Aviles, who set up base nearby at St. Augustine and proceeded to attack the French fleet. Ribaut fought back, but when a hurricane wrecked his ships and stranded the five hundred survivors near the site of present-day Daytona, the fate of Protestant Fort Caroline was sealed.

Through subtropical swamps and undergrowth, Menendez marched overland in another of those feats of derring-do that made the conquistadores the most feared race of conquerors since the Vikings. He attacked Fort Caroline at dawn on September 20, 1565, and put its defenders—men, women and children—to the sword. Over the few he hanged was left this inscription: "I do this not as to Frenchmen but as to Lutherans." A good soldier, he then mopped up the shipwreck survivors, to whom he first promised mercy and then summarily dispatched.

Ribaut died too, his name to survive chiefly on a bowling alley, a strip road and a shopping center in nearby Beaufort, South Carolina. For the Spanish, by their own estimate, it was good that he did. As Menendez assessed Ribaut for his master, Philip II of Spain: "I think it great good fortune that this man be dead, for the King of France could accomplish more with him and 50,000 ducats than with other men and 500,000 ducats; and he could do more in one year than another in ten." The French felt the thrill of revenge, though fleeting, in 1568, when Dominique de Gourgues sacked San Mateo (as the Spanish had renamed Fort Caroline) and hanged his Spanish prisoners on the very spot of Menendez's executions three years before. His cautionary inscription read: "I do this not as to Spaniards, nor as to mariners, but as to traitors, robbers, and murderers."

Bold and butchering gestures aside, the Spanish and not the French wrote the next chapters in the history of this coast. In 1566 they had landed on Parris Island with 150 men. Juan Pardo set out to explore the interior, and Menendez established his capital city of Santa Elena somewhere near the site of what had been Charlesfort. Eventually five hundred people, including women and children, settled here or tried to. Hostile Indians drove them out in 1576; they returned a year later, redubbed the place San Marcos and stuck with it until another enemy, this time a white one (the English), forced their permanent retreat to St. Augustine.

In 1586 a marauding Sir Francis Drake burned St. Augustine (though he did not slaughter its inhabitants) and spread panic in the exposed settlement to the north. The English did not return to Port Royal for good until well into the next century and thus well past the age of discovery, but their intrusion exposed the weakness of Spain's hold on these reaches north of the Caribbean. Proof of that weakness would come later, when Barbados and Jamaica were forcibly switched from Spain's column to England's, and when desire for permanent settlement had replaced the thrill of treasure-seeking as the prime mover driving adventures in the Americas.

That much is already in the history books. But it is how the brief story of Santa Elena has been remembered that requires the journey to Parris Island. Consider those three historical markers out there among the lizards and live oaks at the edge of the Marines' golf course. They relate not only a history but also a historiography that illustrates the uses to which obscure tales of discovery can be put.

You first cross a small footbridge over a dry moat leading into what was once a fort. In the center rises a monument to Jean Ribaut and his ill-fated French Huguenots. A stumpy column, it bears two inscriptions. One, carved under a graphic likeness of an open Bible and the heading Religion, Freedom, Truth, reads: "Here stood Charlesfort, built 1562 by Jean Ribaut for Admiral Coligny, a refuge for Huguenots and to the glory of France." The other, under the crown and fleur-de-lis of the French kings, reads: "Erected 1925 by the government of the United States of America to mark the first stronghold of France on this continent." The only problem is that the blocks marking the fort's outline are now known to encompass not a French but a Spanish site. The up-to-date self-guided driving tour provided by the Marines forthrightly points this out, though it seems likely that the casual visitor who more or less stumbles onto the site will ignore the guide and read just the inscription. ("Ribaut Monument" is what the place is called on printed directions to the site, not "Santa Elena.")

The evidentiary trail of what was really what goes back to the nineteenth century, and my understanding of it has been aided by a report of the University of South Carolina Institute of Archaeology and Anthropology kindly supplied me by Dr. Steven Wise, curator of the Parris Island Museum. Conspicuous surface mounds first attracted the attention in the 1850s of Captain George Parson Elliott, Commander Matthew Fontaine Murray (the famous hydrographer) and William Gilmore Simms (South Carolina man of letters and author of the early novel *Yemassee*), who together made measurements and excavated a gate to what they believed was the old French settlement of Charlesfort. Thinking they were only of Civil War vintage, the Marine Corps leveled the mounds in 1918 to make way for new training facilities, but not before Colonel John Millis, the commandant of Parris Island, made drawings and photographs. The consensus was growing that the remains were indeed those of Charlesfort, and when the results of the major excavation of Major George H. Osterhout were published in the *Marine Corps Gazette* in 1923, the case seemed conclusive that Ribaut's French bastion had been found.

Osterhout's dig unearthed cedar stockade posts, fortification ditches, bastions, and a good haul of artifacts. Close on the heels of Osterhout's news, the Huguenot Society of South Carolina in 1925 commissioned the Ribaut Monument and placed it square in the center of the plot everyone now knew was Charlesfort. It was just after the Great War in Europe (in America, Woodrow Wilson's war to make the world safe for democracy), which had

This is a 1564 drawing of the plan of Fort Caroline, which was built on the St. John's River in Florida. (John Carter Brown Library, Brown University)

witnessed the United States repay its old Revolutionary War debt to France. French stock in America then was high, and besides, then-commandant of the Marine Corps Walter Lejeune was of French descent and hailed from nearby Beaufort. So the happy words about the glory of France appearing on the column flowed naturally enough.

They flowed just in time, because over the next two years conflicting views were published, identifying the site as Spanish, not French. Inadequate knowledge about sixteenth-century artifacts caused the confusion to persist for another thirty years until 1957, when a National Park Service historian, Albert Manucy, re-examined everything from the Osterhout dig and firmly declared all of it to be Spanish.

And so vanished Charlesfort, leaving only the noble monument. The true site is still believed to be somewhere on Parris Island. Exactly where we will probably never know. The archaeologists continue their patient digging, sifting and comparing as the grant money permits. Who knows? One day they may stumble upon a shard of indisputably French provenance and so rekindle the search for Ribaut's long-vanished Huguenot refuge.

The second marker, across the bridge on the other side of the creek bed, relates the tale of early Parris Island in a different style. It was erected by the Beaufort County Historical Society, and its measured phrases that tell you just what happened here (and what didn't) reflect a later era's enthusiasm for objectivity in history and the care for not reaching beyond what the hard evidence will support. It tells of the arrival here on August 18, 1521, of Spaniards from Santo Domingo and of their subsequent frustrating efforts over the next sixty years to plant a permanent settlement. Spurred by the arrival of Ribaut's French Protestants, the Spanish soon

Shipbuilding was very much a part of early seafaring and exploration, as shown in a modern painting by Sidney King. (U.S. National Park Service, Colonial National Historical Park)

This detailed and very decorative 1555 map by Le Testu was made for Gaspard de Coligny, who became the leader of the Huguenots and supporter of colonizing expeditions to Florida in 1562 and 1564. (Musée de la Marine, Paris)

destroyed the heretics, and in the early 1580s their city of Santa Elena boasted sixty-odd buildings and a partially self-governing population of men, women and children. Only after Sir Francis Drake's raid on St. Augustine in 1587 were the inhabitants of Santa Elena withdrawn and the place abandoned.

The marker is brisk, to the point, correct in the essential details. All the important groups—the Spanish, the French, the English and yes, the Indians—get a mention. The key individual characters—Menendez, Ribaut, Drake—take their bows. It is very professional history: fair, balanced, brief.

Today, it and the old French column are joined by a third contender, this one put up by the South Carolina legislature through the South Carolina Institute for Archaeology and Anthropology at the University of South Carolina. Two things visually distinguish it from the others with which it shares this lovely place. For this is not your historic sites agency standard-issue slab of steel or concrete, but rather a colorful ceramic tile triptych-like affair of the sort which, were it merely decorative, you might expect to find in the foyer of an upmarket Mexican restaurant. And as with the menu in such an eatery, you are invited here to read the offerings in English and Spanish too. This one reads, proudly: "In respectful tribute to those Spaniards who left their mark here between 1566 and 1587 while in quest of their country's glory and in grateful recognition of the distinguished Americans who today with their work pay homage to the memory of those heroes and the history shared by the two nations, Spain and the United States. Santa Elena, which is this place, the twelfth day of October, 1982, Hispanic Heritage Day." October the twelfth, when I was growing up, meant Columbus Day. But neither his name, nor the name of any other individual, appears on this cairn, which is a tribute to group heritage, not individual heroism.

As if anyone today cares whether it was on this spit of land or that where Europeans first made their mark, or whether they were French or Spanish. (Parris Island is not heavily promoted by the state tourism authorities, who have put their money on the charm of Charleston.) Yet the makers of the markers on Parris Island cared, and they brought to their task of commemoration notions shaped by the world they themselves—not the discoverers—lived in. The Ribaut Column and the Spanish Heritage triptych most clearly demonstrate the tendency. Our heroic French battlefield comrades must have had—here—heroic origins. The Spanish, derivatives of whose culture today vie for official status in the South, the West and the northern cities of the United States, deserve fuller historical recognition. And at Parris Island they get it, tiles and all.

Both monuments strive to connect the present with the past, to look back and discover or reaffirm their "roots." In distinction, that other marker put up by the local historical society, whose wording is so careful and correct, strives to connect the past with the present. The little protocity of Santa Elena, it tells us, was

ALTHOUGH THEIR EMPIRE ENCOMPASSED SOUTHERN
and Central America and Mexico, the Spanish also forayed into the interior
of western North America before the English or the French got even
a foothold on its eastern edge.

Beginning in the late 1520s, Spanish conquistadores made several rambling expeditions to the interior, hoping to find there riches in gold and
silver such as they had come upon farther south among the Aztec and
the Inca. Disappointment awaited them. Panfilo de Narvaez explored
the unknown country along the Gulf Coast between Florida and Mexico
in 1528. He started near Tampa Bay, but after an overland march to
Apalachee Bay he failed to rendezvous as planned with his ships, and
then set out on a disastrous land and water trek past the mouth of the
Mississippi to the site of Galveston, Texas. Six hundred men had started
out; by the end of the winter fifteen remained alive and they had resorted
to cannibalism.

In 1539, Hernando de Soto set out on his much more famous three-
year expedition into the interior of the southeastern section of the future
United States. He was looking for gold and silver, and from Apalachee

Inland Expeditions

*The confrontation of Renaissance Europe-
ans and Stone Age Americans, while not
always so horrific as depicted here between
the Spanish and the Indians in Theodor
de Bry's* America *from 1596, resulted in
the decimation of native populations and
the triumph of European ways in the
American wilderness. (New York Public
Library, Astor, Lenox and Tilden
Foundations)*

Bay struck out to the northeast through the pine forests of present-day
Georgia to the Savannah River. Encouraged by the Indians' tales of riches
farther on, they marched northwest into the Appalachian Mountains
and then south to winter near Mobile Bay. Their relations with the Indians
disintegrated: 2500 Indians were slaughtered at Mauvilla. The Spaniards
and their fierce herds of dogs then zigzagged north and west across the
Mississippi and Arkansas rivers onto the Ozark Plateau in search of a
path to the western ocean, which they did not find. Returning toward
the Gulf, de Soto died; the remnants of the expedition floated in makeshift
boats down the Mississippi and, miraculously, reached Mexico in
September of 1542.

While de Soto's men were still wandering about the Southeast, Francisco
Vasquez de Coronado explored northward from the Gulf of California
into the barren reaches of Arizona and discovered the Grand Canyon
of the Colorado River. Searching for legendary cities of gold, he found
only Indian pueblos and to the east the immensity of the Great Plains
in what would become Texas, Oklahoma and Kansas. He marveled at
teeming herds of buffalo and other wildlife but looked in vain for the
riches that drove the conquistadores to such epic treks.

partially self-governing and so furnishes us today with an early example of values and behavior that in our own times have received universal approbation: "Use of the democratic process and of women's suffrage." It's a bit of a stretch, the Spanish dons never having been famous for fostering political freedom in the foreign lands they ruled. But there you have it: this site commands attention today because from what was done here long ago mighty things later grew.

Drive back now, past the golf course and Page Field, to the Parris Island Museum, where you will witness the same oddness, the same warp of two different times bearing down on one small place. Sign in with the young Marine on duty (soft duty) at the desk, check your attire against the dress code (no bare feet, bathing suits, cut-offs, or clothes deemed to be obscene), move past the rooms filled with case after case of Marine Corps regalia from every era of their history, and find the small alcove devoted to Parris Island's pre–Marine Corps past.

There are cases filled with artifacts from the various digs, maps, dioramas, and a larger than life-sized mannikin, dressed in French military costume of the style worn by Jean Ribaut and his officers on formal occasions, about whom the curator seemed a bit embarrassed. There is also a kind of timeline showing the way history, one might say, piles up. Its different strata reveal a Low Country archaeological profile and contain artifacts of progressively later human occupation.

Four cheery murals depict Spanish life here, including a cozy domestic scene with father, children and wife in the background; a religious procession led by two altar boys; a priest celebrating mass against a backdrop of the sea; conquistadores treating peacefully with the natives. Relics abound, spanning the ages: lead shot from the Plantation period; a hand-blown bottle; Spanish pottery shards, and a fragment from an olive jar that has been rounded off into a token for the game of checkers.

Read all the labels and you will get the message that they really don't know where Charlesfort was, that the French were beaten off by the Spanish, who stayed until they were frightened off by the English. The history of European contact then undergoes an eight-year hiatus until English sea captain William Hilton, sailing from Barbados in 1663 for the Lord Proprietors of the newly created colony of Carolina (after Charles II of England), entered Port Royal Sound and examined the ruins of Santa Elena (or was it Charlesfort?) and its fortifications. Three years later English colonists attempted their own settlement here, but were turned away because of stories of intertribal warfare and the continued proximity of the Spanish. Instead they went north and founded what would become the city of Charleston, which forever after overshadowed backwater Port Royal. Hilton did, however, leave his name behind (Hilton Head), and that—not Ribaut or Menendez—is the name

known by every golfer, resort fancier and Marine in the region. What all of these men did, the displays seem to say, relates to a more recent past and to more familiar national experiences. Somehow we are made to feel the obligation to pay them homage for coming first, for making the landfall, for blazing the trail.

And it was the English, under Sir Francis Drake, who frightened off the Spanish from this coast and so sealed the fate of Santa Elena. But these were English of another order from those of modern times: a Gladstone, a Disraeli, a Churchill, a Thatcher. They were Elizabethans—Shakespeare's contemporaries—whose understanding of freedom and the social order differed vastly from our own and even from that of our Founding Fathers two hundred years ago. Religion, not ideology, moved their age, and the religious settlement worked out at home by their sovereign Elizabeth I between extremes of reform and reaction left most of them reasonably content in matters to do with faith and politics. Drake, whose appearance along this coast in 1586 ousted the Spanish from Parris Island for the last time, introduces for us this race, who in North America emphatically did succeed in making discovery permanent. But for them as for all the others, it was always a close-run thing; their motives and efforts ever demand from us some leap of historical imagination to appreciate. The place they first tried was Roanoke Island at the very moment that Drake dispatched the dons from Port Royal, where four hundred and some years after that the United States Marines still play golf near their graves.

Throughout the history of discovery, the factual and the fanciful combined to form Europe's image of the new adventure. Engraver Theodor de Bry captured the spirit in this 1594 rendering of stout galleons afloat in a sea of flying fish. (New York Public Library, Astor, Lenox and Tilden Foundations)

Roanoke Patriots

FOUR HUNDRED MILES NORTHWARD, UP THE Atlantic Coast, Roanoke Island faces a shallow sound. In the 1520s Giovanni da Verrazzano mistook this sound (actually there are four sounds surrounding the island: Roanoke, Croatoan, Pamlico and Albemarle) for the Pacific Ocean, an observation that, had it proved correct, would have reduced the North American continent to a strip of land a mile wide. It was an understandable error, for from the masthead of a sixteenth-century ship you could not see across it. These are large but strangely limited branches of the sea. Shallow water permits transit only by pleasure craft and the custom-designed shallow-draft ferries that the State of North Carolina operates between the mainland and the islands of the Outer Banks, and among those islands themselves. Four hundred years ago, the only traffic here was Indian canoes and, for two or three years in the late 1580s, an occasional English pinnace and longboat.

The sounds are separated from the Atlantic Ocean by a string of barrier islands reaching from the Virginia state line 350 miles south and southwest around Cape Hatteras and Cape Lookout to Cape Fear. While shallow and accessible from the ocean only via constantly shifting and even shallower inlets, the sounds are known to churn up a respectable chop in a northerly gale. They are at once self-contained and exposed; hemmed in by the land and only a breath away from the Atlantic in its wildest reaches. Mariners have long known this coast as the Graveyard of the Atlantic; its great shifting arms of sand—Frying Pan, Diamond, Wimble shoals—have sprung the seams and broken the backs of hundreds of unwary vessels, from the little oaken barks of Queen Elizabeth I's day to the mammoth steel steamers of our own. Compare the view with the one from Cape Henry at the mouth of Chesapeake Bay, which is filled with warships and merchantmen. From the shore here, at Kitty Hawk or Kill Devil Hills or

Walter Raleigh outlived both the Roanoke venture and Elizabeth I, his great patron. Known as "the last Elizabethan," he symbolized the English fascination with the New World on the eve of settlement. (National Gallery of Ireland)

Map of Eastern North America: Florida to Chesapeake Bay by John White. The coat of arms is Sir Walter Raleigh's. (British Museum)

Nags Head or Jockeys Ridge, peer eastward—toward Gibraltar—and behold only ocean. Sailors steer far out to sea to avoid this coast, which offers no welcoming harbors and where no great cities grew.

That such an unlikely spot occupies such an important place in the history of discovery is owing to a peculiar set of circumstances and motives, miscalculations and bad luck some four centuries ago. It is also owing to a deliberate commemorative act some fifty years ago, by which Americans have come to revere this spot out of all proportion to the effect it actually had on history. What happened here—the attempt of Elizabethan English men and women to establish a permanent settlement based on trade and farming—failed. What we know of their efforts along this coast in the last half of the 1580s is not inconsiderable. The episode of Roanoke Island in fact gave Europe some of its earliest and most memorable reporting about conditions in North America on the eve of settlement. Yet what we do not know outweighs even this: the English on Roanoke Island who were seen by other English in 1587 were never seen by their countrymen again. Their fate is not known—except that they were somehow swallowed up by America. Here, history cannot answer its own most basic question of what finally happened, though it makes some guesses about how it might have happened. And where facts fail, fiction begins.

Of all the facts that we have about this chapter in the history of discovery the most amazing are not in the written record at all. We know much of what the "Lost Colony" of Roanoke looked like, and we know it with a precision and see in it a beauty that is amazing before the age of photography. The Roanoke episode has come down to us illustrated.

Drawings of Roanoke by artist John White, who also was governor of the colony, have survived in the British Museum and are today handsomely reproduced and widely available through the cooperation of the museum, and University of North Carolina Press, and the Four Hundredth Anniversary Committee of the North Carolina Department of Cultural Resources. Even though the great bulk of White's work was lost to the sea when he departed Roanoke for the first time in 1586, what remains constitutes the most detailed visual rendering we shall ever have of one part of this continent—and one group of the people who already lived

Indians dancing by John White. (British Museum)

here—at the moment of European intrusion. I summarize from the description of Paul Hulton, formerly deputy keeper in the Department of Prints and Drawings of the British Museum.

The town of Pomeiooc, a palisaded Indian village on the mainland near Mattamuskeet Lake, with pole and mat longhouses furnished with interior platforms used mainly for sleeping; Indians gathered round a central fire; a man chopping timber with an ax. The Indian village of Secoton, an open town of thirteen houses and much activity, including Indians dancing around a post, figures squatting to eat, a charnel house, fields of corn in three stages of growth ("newly sprung, greene, and rype"), a scarecrow platform. An Indian woman and young girl, the woman probably wife of a chief, her hair fringed, a headband of beadwork, shell necklace, her face painted or tattooed, the prepubescent girl wearing only a moss patch at the crotch and a bead necklace. An old Indian man, hair knotted at the back, clad in a fringed deerskin. Indians, ten men and seven women, two of the men in breech-clouts, the others in apronlike skirts, carry gourd rattles, leafy sticks and spears, dancing around three women who are embracing each other in the center of a circle. Indians around a fire: six men and four women gathered about a fire with pumpkin rattles, some clothed in skin mantles, some naked. A man and a woman eating: two figures squat on a mat around a circular pan of hominy or maize. An Indian priest of Secoton, clad in special vestments of rabbit skin and leather ear ornaments. Fishing: a composite of four natives in a dugout canoe filled with fish and a fire in the middle, for fishing at night, spears, dipnets and fish-weirs; catfish, hermit and king crabs, sturgeon and hammerhead shark fill the water while wildfowl fill the skies overhead. A cooking pot, built up in coils by hand without a wheel, with an ear of corn visible at the surface of a stew likely containing a variety of fish, flesh and fruit. Cooking fish: fresh fish, probably shad, being broiled, not dried, on a barbecue frame made of sticks over a spoked fire. An Indian in body paint: a figure with a longbow adorned for some special occasion. An Indian conjuror: distinct from a priest, with a small bird worn at the side of the head as a badge of office and a pouch commonly worn at the waist to hold pipes and tobacco. There are also fish and flowers (other sorts of plants, and the pictures of mammals, seem to have been lost): land crab, hermit crab, fireflies, gadflies, scorpions, milkweed, the head of a brown pelican, burrfish, loggerhead turtle, common box tortoise, diamond-back terrapin, swallowtail butterfly, flying fish, soldier fish and grunts.

Altogether about seventy-five watercolors have survived. They are the work of a skilled limner, charged to record as fully as possible the human and natural life in a startlingly new geographical setting. Brought back to England by White in 1586, the drawings were, through the offices of geographer Richard Hakluyt and Sir Walter Raleigh, promptly and expertly engraved on copper by Theodor de Bry, a native of Flanders and a Protestant refugee in Frankfurt-am-Main. It was de Bry who in 1590 published Thomas

Scientist and mathematician Thomas Hariot, together with John White, left a remarkable series of drawings of the New World's flora and fauna. (British Museum)

Hariot's *Briefe and True Report of the New Found Land of Virginia* in Latin, English, French and German editions. (Hariot was White's scientific companion at Roanoke.) De Bry went on to produce many volumes in his *America* series (the second was Jacques Le Moyne's account of René Laudonnière's ill-fated Huguenot colony in Florida in the mid-1560s), but none achieved the same impact of White's images of the southeastern Algonquians along the coast and on the islands of present-day North Carolina.

For three hundred years, knowledge of the existence of the drawings was limited to specialists in art and natural history, until the first facsimiles were exhibited on the occasion of the nineteenth century's great commemoration of Columbus's American discovery: The World's Columbian Exposition in Chicago in 1893. (They were platinum photographic prints, hand-colored at the Smithsonian Institution in Washington, D.C.) The originals were first exhibited outside England in Raleigh, North Carolina, Washington and New York City in 1965. Today, they are widely known and rightly touted as one of the greatest legacies of the Roanoke colonies, the likes of which Jamestown and the Pilgrims' Plymouth cannot begin to match.

As startling as the images no doubt were when published in late sixteenth-century Europe, they are still startling today. Then, they conveyed all the mystery that surrounded an impossibly far-away place whose significance was just about to burst onto the consciousness of a ready audience. Now, they convey all the mystery of a faraway time, when not only what men saw but how they saw it was vastly different and somehow odd to modern eyes. For the purposes of constructing a historical narrative, we regret that among White's drawings there is no depiction of the English themselves at Roanoke. Recording what the English already were familiar with was not indeed White's commission, and he must have been a very busy man just sketching the natural world, which, for "marketing purposes," needed to be handsomely and promptly reported back home. But the absence of such images of the intruders—and the presence of vivid likenesses of the natives who were first their hosts and then their adversaries—is a help at least in this regard: it enables us to see, almost literally as it were and without distraction, the new world through sixteenth-century eyes. For these strange-looking people, drawn by a good but certainly far from great artist, in a perspective that never seems quite right, were to the arriving Europeans truly odd beyond measure.

The Indians had no writing and left no record of themselves for themselves or for anyone else, and Hariot's words must be read with his motives in mind. He recounted the Indians' clever wit, their fascination with English tools and their joining in the English prayers, but he glossed over their warlike characteristics and said nothing of the Indians turning against the English or of the settlers' betrayal by their chief. Hariot portrayed them in as unfrightening a light as possible: as fit companions for the English who were about to venture off to the New World. Yet

Walter Raleigh, favorite of Elizabeth I and founder of the ill-fated Roanoke colony. (National Archives of Canada)

there also seems to be genuine charity in his judgments. The thought of coexistence, however fleeting or ingenuous, was not, at this earliest moment, necessarily naive. If ever it had a chance, it was here at Roanoke where the slate was clean.

What the natives did—the sort of lives they led—made them easy game for men who acted on the world differently. Far from savages (though that was the term the English commonly used to describe them) like the Arawaks, whom the Spanish came upon in the Caribbean and quickly exterminated, they were a social people tied together loosely in tribes and families, whose development still fell short of writing, the wheel, fashioning metal into tools, the use of money and the market. They had no machines, even as the ancients had known, nothing in any way to leverage their own physical strength against the forces of nature. So they lived remarkably at peace with their surroundings, whatever they happened to be. They had no notion of "in-doors" and "out-doors" or that their roofs had any significance beyond keeping out the rain. They had learned many of the secrets to unlocking nature's crude bounty, and they enjoyed it fully at those times and seasons when nature saw fit to dispense it. But they did not know how to and were not inclined to go nature one better. Though they raised corn, they did not "farm" in the European sense of setting by stores against the cold months following the harvest, or against bad years to come. Seasonal subsistence, not surplus, was the aim of their agriculture. They ate well in the late summer and autumn, much less well in the spring.

This explains the usually catastrophic effect that even small numbers of Europeans could have when they looked to the at first friendly natives for food. Propagandists for colonization like Hariot (scientist though he was) commonly praised the food-giving potential of the new land and reported the bounty they had experienced at Indian feasts. But the Indians lived precariously, ever on the edge of famine. They knew how to reap nature's edible gifts, but they could do it only for themselves and never sufficiently at that. They stayed alive by the skin of their teeth, according to nature's rhythms, with no thought (at least none that was discernible to outsiders then and certainly is not to us now) that tomorrow would or should be any different from today.

Indians fishing by John White. (British Museum)

In Raleigh's great American venture, Elizabeth I backed him with both spirit and substance. Though she never permitted her great favorite to go there, Raleigh named the place after her virtue: Virginia. (National Maritime Museum, Greenwich)

About those Englishmen who are not in the pictures, the remaining facts come to us as words on paper (and in a few portraits painted at home), and for the sixteenth century they are relatively abundant. They document the first serious effort of the English to take hold of the claim to North America that John Cabot had made in the name of Henry VII in 1497. "The Roanoke Voyages" is how the episode is commonly called. It is the story of how between 1584 and 1590 men, and then men, women and children, from the west of England and Ireland set out to extend the realm of Gloriana in the New World and, with luck, to do them-

selves a good turn too. With the nature-bound natives whom they first confronted, there could hardly have been greater cultural contrast, or less hope of happy relation. For unlike the Algonquians who appeared to do nothing (and what they did do, they seemed to do over and over again), the Raleighs, Grenvilles, Lanes, Hariots and Whites of the Roanoke Voyages burst onto the American stage fully formed by the English Renaissance, ready to do everything.

As things appeared from England at least, the time was right for such a venture, the foundations well prepared. A hundred years before, frugal Henry Tudor had ended the Wars of the Roses, united England and forged the beginnings of a modern monarchy. He passed on to his son, Henry VIII, a full treasury and a high intelligence, and in a tumultuous reign the son spent, but did not squander, his inheritance. Domestic and European preoccupations kept his attention and energy away from New World vistas, and he did not follow up Cabot's discovery. But two of his policies prepared the way and provided the context for later efforts that did. Henry's break with the Pope over annulment of his marriage to Catherine of Aragon set England fitfully but decisively onto the Protestant path. During a century when the energies released by Luther's Reformation split the Continent into armed camps and spilt the blood of thousands, England under the Tudors managed to avoid religious war, even as Henry despoiled the Church in England of vast wealth and boldly took the Pope's place at the head of the new Church of England. At the same time, he enhanced his kingdom's place amid the powers of Catholic Europe by building a modern Royal Navy fit to command the Channel—the Narrow Seas—that separated England from her enemies. Two of his three children to succeed him (each mothered by a different queen)—the Protestant Edward VI, a sickly boy, and the Catholic Mary, who excelled only at making Protestants into martyrs—advanced their inheritance hardly at all. But by the time of Mary's death and the accession of her sister, Elizabeth, in 1558, one thing was growing plainer. England could not forever avoid conflict with a Continent in the throes of counter-reformation, or with the country that most fiercely represented Catholic reaction: the Spain of Philip II.

The early years of Elizabeth's reign were filled with cautious maneuver, the plot and counterplot that secured her insecure throne, and by a diplomatic dance with Spain that kept peace even as the interests of the two kingdoms increasingly diverged. Overseas, Spain had consolidated a vast colonial empire in South and Central America and as far north as Florida, which every year sent treasure galleons laden with the wealth of the Aztec and the Inca eastward across the Atlantic. Philip was no Henry Tudor, and an ambitious foreign policy in Europe put even his newfound American wealth under strain. When Elizabeth, who sensed well the profound Protestantism of her subjects, rebuffed his offers of marriage and dynastic alliance, Philip planned extravagantly to invade England. War came, and in 1588 his Armada sailed.

The English, however, had waged a de facto war at sea against the Spanish since the 1570s. It was an unofficial but royally condoned hit-and-run war on Spanish shipping by England's first great captains: John Hawkins, Francis Drake, Richard Grenville. Part pirates, part patriots, they harassed the Spaniards at will and enriched themselves—and Elizabeth—with the booty. In the early days of this free-for-all, when Spain was unprepared, it had been easy enough to pick up handsome prizes as the treasure fleets approached the coast of Spain after the long voyage from the Caribbean. English guns and crews were fresh, and it was only a short voyage home with the loot. It was too good to last, and when the Spanish took action to protect their western approaches, Elizabeth's "sea dogs" (as legend romantically knows them) were driven to the far side of the Atlantic to await their prey in remoter waters: the Florida Straits and the other Caribbean chokepoints that the galleons had to pass en route from Panama, Cartagena and Havana, up into the Gulf Stream and on toward home. The distance was daunting and considerably raised the risks for the English pirates.

The lack of a friendly harbor within three thousand miles spurred them to think about finding one, and the attempt to establish a colony on Roanoke Island is a direct result of that need. To increase the chances of successful plunder, an English fleet based in America, its guns and crews unfatigued by the long voyage across the Atlantic, seemed just the thing. In the men who tried it, motives of profit, propagation of the Protestant religion, and sheer buccaneering adventure, all played their parts. Plus there was patriotism, in the form of a vision of English empire beyond the seas. For a hundred years the cautious English had watched the wealth of New Spain enrich their greatest rival. Now at last they seemed ready and able to do something about it.

The first move toward establishing a multipurpose American outpost belonged to Sir Humphrey Gilbert, who received a charter from Elizabeth in 1578 to discover and settle in the New World. He sailed to St. John's, Newfoundland, in 1583, stayed only a month, and then on the return voyage in September "was swallowed up by the sea." In Sir Humphrey's failure, Walter Raleigh, his half-brother, saw opportunity and did not hesitate. Then the great court favorite of his queen, Raleigh entered American history like a man in a hurry, which he was. Within a month of receiving his own patent, he had prepared an exploratory expedition of two small ships under the command of two gentlemen from his London and Devonshire households, Philip Amadas and Arthur Barlowe. They sailed with clear instructions, good instruments and the best maps of the times. The maps were inspired by a Welsh magician, John Dee, who had his information from a renegade Portuguese pilot, Simon Ferdinando, who had surveyed the American coast for Gilbert in 1580 and had reported, about halfway between the West Indies and the Newfoundland Grand Banks, a fine bay lying behind a string of islands. It was then called by the Spanish Bahia

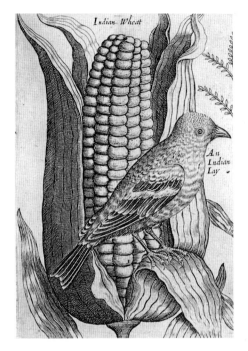

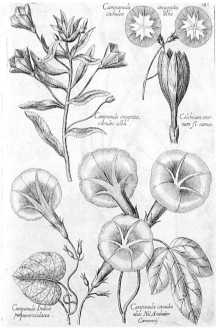

American flora and fauna continued to fascinate Europeans for decades after discovery was secure. Here, an illustration of a redwing blackbird from Edward Bland's Discovery of New Britaine *(1651), and a morning glory from Theodor de Bry's* Florilegium *(1641). (Courtesy of the Trustees of the British Library)*

Hermit crab and swallowtail butterfly by John White. (British Museum)

de Santa Maria, and it described the sounds behind North Carolina's Outer Banks in the midst of which lies Roanoke Island.

Falling between 35 and 37 degrees north latitude, the area lay within the climate range of the Mediterranean, and it seemed to Raleigh the perfect nest for pirates (his pirates) and a promising place for a settlement. With Ferdinando as pilot, the two small barks slipped out of Plymouth unnoticed by the Spanish in April 1584, took the southern, tradewind route via the Canary Islands and the Caribbean and arrived off the Carolina capes, probably between Cape Fear and Cape Lookout, on July 4. Then they coasted north for nine days along the Outer Banks. Barlowe caught the character of the place: "We sawe before us another mightie long Sea: for ther lieth along the coast a tract of Islands, two hundredth miles in length, adjoyining to the Ocean sea, and between the Islands, two or three entrances: when you entered betweene them, (these Islands being very narrow . . . in most places sixe miles broad, in some places less . . .) then there approacheth another great Sea fortie, and in some fiftie, in some twentie miles over, before you come unto the continent and in this inclosed Sea, there are about a hundredth Islands of divers bignesses." Somewhere in the "inclosed Sea" was where they wanted to be.

On July 13 they found their way to it through a shallow opening in the sand barrier near present-day Oregon Inlet, which they named Port Ferdinando after their pilot. They crossed into Pamlico Sound off Hatarask Island at high tide, with just twelve feet of water under the keels of their little fifty-ton barks; they were about ten miles south of Kitty Hawk and the great sand dunes at Nags Head. After three days the Indians appeared with offerings of wine, meat and fish, and from them the English learned that their "werowance," or chief, was called Wingina and that he ruled the Roanoke tribe from a village on nearby Roanoke Island. The Indians' name for the region (which did not stick) was Windgandcon.

Barlowe and seven of his crew took a longboat and explored north, twenty miles into Albemarle Sound, and visited Wingina's village on Roanoke Island, where they were entertained in a manner that, as Barlowe reported it, was sure to confirm the enthusiasm of Raleigh, his patron: "Sodden [stewed with corn and beans] Venison, and rosted, fishe sodden, boyled, roasted, Melons rawe, and sodden, rootes of divers kindes, and divers fruites: their drinke is commonly water, but while the grape lasteth, they drink wine, and for want of caskes to keepe it al the yeere after, they drinke water, but it is sodden with Ginger in it, and blacke Sinamon, and sometimes Sassaphras, and divers other wholesome, and medicinable hearbes and trees. We were entertained with all love, and kindnes, and with as much bountie, after their manner, as they could possibly devise. Wee found the people most gentle, loving, and faithful, void of all guile, and treason, and such as lived after the manner of the golden age. The earth bringeth foorth all things in aboundance, as in the first creation, without toile or labour. The people only care to defend them selves from the cold, in

Indian village of Secotan by John White. (British Museum)

theire short winter, and to feede themselves with such meate as the soile affoordeth: their meate is very well sodden, and they make broth very sweete, and savorie: their vessels are earthen pots, very large, white, and sweete: their dishes are woodden platters of sweete timber: within the place where they feede, was their lodging, and within that their Idoll, which they worship, of which they speake incredible things." On the other hand, Barlow also told of Indian clubs fashioned from sharpened stags' horns and clearly sufficient to kill a man, and of wars "very cruell and bloodie, by reason whereof, and of their civil dissentions, which have happened of late years amongst them, the people are marvelously wasted, and in some places, the Countrey left desolate."

They did not linger and, their reconnaissance made, set sail for England on August 23. A fast passage landed them at home in just three weeks. To their report, Barlowe appended his and Amadas's names, along with those of eight of the ship's company, so as (he explained) to strengthen the claim of possession they had made in the queen's name. And then he added, as if it were a postscript, this note: "We brought home also two of the Savages being lustie men, whose names were Wanchese and Manteo." We must presume they came voluntarily. Without one of them, Manteo, the next expeditions would not have lasted even as long as they did.

The next year, 1585, the Roanoke effort grew in scale and intensity. We cannot know if the voyage of that year and the subsequent one of 1587 would have had happier results if Walter Raleigh himself had gone along. We know only that he didn't, and for the very good reason that his queen would not let him. He was her favorite, and lover in all but the physical sense, and she would not let him far from her side. He must, she said, save himself for England's defence, which she correctly saw growing ever more urgent. Truth be known, if the dashing courtier had had to decide for himself between the chance of battling the Armada up the Channel alongside Sir Francis Drake (which in fact he did, with distinction), and adventuring off to the New World, he might have been hard-pressed to make the choice. Luckily for Raleigh, Elizabeth saved him for herself and history's greater glory. The men who went in his stead and at his favor on the next trip out were, like him, brash soldiers and adventurers eager to make quick fortunes and return to enjoy them in England. They were brave enough characters, and their actions mirrored their motives. Sadly, those actions did little to lay foundations for the sort of lasting settlement that would depend for its survival on the goodwill of the native peoples. This time, Raleigh's surrogates left behind in the new land of Windgandcon enough blood and bad feeling with the Indians that the third and final Roanoke expedition of 1587, which included women and children and was intended for permanence, was "lost" before it even began.

This second follow-up foray was very much a military affair, faithful to Raleigh's foremost privateering instincts. With Barlowe's report in hand, he had attempted to persuade Elizabeth to elevate his colonizing schemes to more official status, but the canny queen would still not go the whole way into this American venture, even for Walter. She did, however, bestow a knighthood on him at Epiphany, 1585, plus a lucrative new court office as Master of the Horse and a fine title: *Walteri Ralegh Militis Domini et Gubernatoris Virginiae* (Walter Raleigh, Knight, Lord and Governor of Virginia). With the title went permission to call the new colony after her honor: thus Windgandcon became, felicitously, Virginia. Elizabeth did lend the expedition a prime ship of the Royal Navy, the *Tyger*, contributed four hundred pounds of gunpowder from the royal magazine and gave Raleigh's captains authority to take up men for the voyage with press-gangs, as if in wartime. With diplomatic relations with Spain now broken and a hot war not far off, Elizabeth also broadly issued letters of marque, turning pirates into privateers, and she braced for the onslaught she knew in her bones was soon to come.

Raleigh knew it too, and for him it meant time was especially short. He wasted none of it. He sent his own ship, *Roebuck*, out in search of prizes to raise money for the enterprise; he built two

Sir Walter Raleigh's seal as Lord and Governor of Virginia has his personal motto, "Amore et virtute." (British Museum)

pinnaces to join his *Dorothie;* he secured the *Elizabeth* along with her captain and future circumnavigator, Thomas Cavendish, as well as a Chichester vessel, the *Red Lion.* Naval command of the expedition went to Raleigh's cousin, experienced soldier of fortune Sir Richard Grenville, a temperamental fighter who had pleaded and failed to get Elizabeth to let him follow in Magellan's track across the Pacific (a job that went instead to Francis Drake). It was Grenville who later, in 1591, famously fought his ship *Revenge* to her and his own death against the whole Spanish fleet off the Azores and so entered British maritime legend. In 1585, Grenville's job was to ferry the adventurers safely to Virginia—and to take prizes en route to help pay expenses and return a profit to Raleigh and his backers. Responsibility for governing the colony once ashore went to Ralph Lane, an equerry to Elizabeth, a shareholder with Raleigh in privateers, and a ruthless veteran of the wars in Ireland where Raleigh was also deeply involved. Simon Ferdinando the pilot sailed again, this time as master of Grenville's flagship. A total of some five hundred men set out, about half of them sailors, 108 classified as colonists, the rest soldiers and "experts." (There were, for instance, two German miners, a Jewish mineral assayer from Prague, artist John White, and mathematician/scientist Thomas Hariot. No women numbered among the company.)

As had Barlowe's exploratory trip the previous year, Grenville's fleet took the southerly route and made stops at Puerto Rico and at Columbus's original colonial capital on Hispaniola, where the English traded peacefully with the intimidated Spanish for cattle to take to Virginia and ginger and tobacco to sell back in England. They then caught the Gulf Stream and reached Virginia on June 25, 1585, when pilot Ferdinando attempted to take the fleet across the bar via an unfamiliar inlet. *Tyger*—the queen's ship—grounded, and though Grenville skillfully managed to get her off again, most of her cargo of stores for the colony was ruined. With most of his ships thus left anchored in the open ocean, Grenville sent a small boat to Roanoke to tell Chief Wingina of the Englishmen's return. He then set out exploring Pamlico Sound with a party that included Lane, White, Hariot, Cavendish, Amadas and the Indian Manteo, who was now bilingual. White successfully made sketches of the two Indian towns Pomeiooc and Secoton, but at the village of Aquascogoc, the theft of a silver cup prompted Grenville rashly to burn the Indians' huts and spoil their corn. They then sailed the main fleet sixty miles up the coast and were invited by Wingina's brother, Granganimeo, to make their colony at Roanoke Island, which was where Amadas and Barlowe had also recommended. By late August, Grenville and the fleet headed back for England, capturing more than enough prizes to please Sir Walter, to whom they also brought the good news that his colony was duly planted.

Lane's men set about building their fort at the north end of the island and exploring northward to the Chesapeake in search of

Sir Richard Grenville, who led the expedition to Virginia, was reputed to be a difficult leader who quarreled with a number of his associates, including Ralph Lane, the governor of the colony, and Simon Ferdinando, the fleet pilot. (British Museum)

Before the Europeans

TO THE EARLY EUROPEAN VOYAGERS WHO LANDED along eastern North America, the native North American peoples who gazed out on the Europeans' strange ships and first settlements represented a Stone Age culture which had occupied this continent already for many centuries. Unlike the great Meso-American civilizations south of the Rio Grande—the Maya, the Toltec, the Aztec and the Incas—which boasted developed agriculture and mining, northern native peoples in what would become the United States and Canada were both less regimented and less advanced.

Indian elder or chief, woman and conjuror by John White. (British Museum)

Sophistication of social organization ranged from the structured society of the Natchez tribe (in present-day Mississippi) to the nomadic hunters and fishermen of Newfoundland and Labrador. Settlement patterns of the Algonquian-speaking tribes that extended from the Carolinas to the St. Lawrence, centered on the stockaded village sometimes protected by bastions and firing platforms. Its architecture varied with the climate: in the north were conical wigwams of birch bark, moose or caribou skins, bark and pole longhouses farther south; and in the Southeast were loose bark-covered houses for the summer months and partially underground hothouses for the winter.

Agriculture permitted a semisedentary way of life and in the Southeast centered on the cultivation of maize or corn, squash and beans. In the north where maize would not grow, the tribes survived by hunting and fishing, and the population density was correspondingly lower. Religious ceremonies centered on an array of spirits and deities associated with the natural world; the corn dance which serves as the opening for the play "The Lost Colony" is a good example.

Indian political organization reflected the fragile nature of a culture where there was no surplus. Families were organized matrilineally, tribes organized families, and on occasion confederacies brought together tribes usually for war-like purposes. Destructive low-level, hit-and-run warfare was endemic, and performance in battle was key to male standing in the tribe.

None of the confederacies approximated European notions of the "state" that the discoverers brought with them, and none was equal for long to resisting the white men's conquest. It is thought that the Indians managed to bring only one percent of North America's arable land under cultivation, and their populations remained accordingly small. At contact, those in all of eastern North America north of Mexico probably numbered just 500,000.

the proper deepwater harbor whose importance the grounding of the *Tyger* had made painfully clear. Lane penetrated Currituck Sound, rounded the headland that would be called Cape Henry into Lynnhaven Bay, and probed up the Roanoke River in search of treasure. The Indians wore copper ornaments whose source was said to be nearby, and the Spanish had reported silver mines in New Mexico but had gotten the longitude so wildly wrong (not unusual in the sixteenth century) that it seemed to the English that El Dorado might await just a few days' march from Roanoke. The old visions of gold—of the storied wealth of the Indies, in one form or another—would not let them go, and in this second Roanoke episode they blinded the English to more practical concerns, like getting on with the Indians and growing their own food. About his compulsive search for a mine, Lane bluntly confessed: "For that the discovery of a good mine, by the goodnesse of God, or a passage to the Southsea, or someway to it, and nothing els can bring this country in request to be inhabited by our nation."

Lane also comprehended that another condition had to be met if a colony here was to take hold, and in this judgment alone he was completely correct: "Provided also, that there be found out a better harborough then yet there is, which must bee to the Northward, if any there be." He had in mind (just as Raleigh did, first and foremost) a good harbor to serve as a protected base for English privateers, who needed to emerge quickly from their hiding place, pounce on their Spanish prey and then disappear. Sheltered and defendable Roanoke might well have been, but accessible for hit-and-run raiders it was not.

Lane's group had arrived too late in the season to harvest for the upcoming winter of 1585, and when expected resupply did not promptly come, the English leaned even harder on the Indians to feed them from the produce of their own fragile agriculture. As they did so, relations between the two races worsened. With amazing celerity in such an allegedly bountiful paradise, famine threatened. Lane dispersed his company to forage their way through the winter at scattered sites along the coast, and left himself with only about forty men at Roanoke. In this weakened condition, he uncovered an Indian conspiracy led by Wingina (who had changed his name to Pemisopan) to surprise and murder the English officers by setting fire to their huts at night and then beating their brains out. Good soldier that he was, Lane did not hesitate to launch a pre-emptive strike of his own. In early June 1586, he surprised Wingina in his mainland village of Dasemunkepeuc and, at the watchword "Christ our victory," opened fire, scattered the natives and beheaded their duplicitous chief.

At that, the direct Indian threat receded, but so too did prospects for long-term cooperation. Yet the long term was not just then of much interest to Lane and his men, for only a few days later relief arrived. It was not Sir Richard Grenville's long-awaited expedition sent by Raleigh that Lane's overjoyed lookouts sighted off

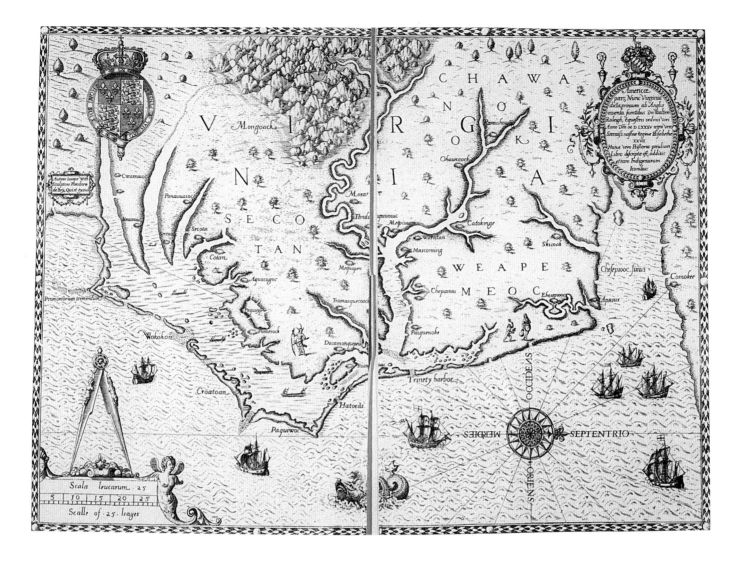

Hatarask Island, but Sir Francis Drake's fleet of twenty-three sail and 2300 mariners returning to England after a rip-roaring raid on the Spanish Caribbean, where they had sacked Cartagena and burned St. Augustine (and frightened off for good the dons from Santa Elena). Ever the gent, Drake generously offered supplies and even a ship for the colony's use in coastal exploration, all of which Lane gladly accepted. But again, the lack of safe harbor intervened to change the intended course of events. A severe storm blew up, lasting four days. Drake's fleet, anchored in the perilous open Atlantic across the Outer Banks from Roanoke Island, was forced to seek sea room. The ship he had offered to Lane, the seventy-ton bark *Francis*, was blown so far that she never came back and proceeded directly to England. Although Drake regrouped and even renewed his offer of assistance with yet another ship, Lane's soldiers of fortune by this time had had enough. They still had no sign that Raleigh's promised resupply under Grenville would ever arrive; supplies were low, and the Indian enemy, if not at the gates, certainly lurked in the woods nearby. So Lane took up Sir Francis's other offer, of transportation for the whole company back to England, and on June 18, 1586, all the English left Virginia.

Scientist and mathematician Thomas Hariot played a vital role in the planning and execution of the Roanoke voyages. With John White, he mapped the new land around its natural inhabitants. His map of Virginia (with west at the top) first appeared in his Brief and True Report of the New Found Land of Virginia *in 1588, the very year the Spanish Armada delayed Raleigh's resupply mission and so sealed the fate of the Lost Colony. (The Newberry Library, Chicago)*

Raleigh had not been as neglectful as it seemed at the time to Lane. He had chartered a two-ship relief force a year before, as soon as Grenville first returned, only to have the queen divert it at the last moment to Newfoundland for a lucrative raid on the foreign fishing fleets working the Grand Banks. A full year passed before another effort could be mounted, and as luck would have it Raleigh's outbound ships (again under Grenville) crossed in the night with Drake's fleet returning Lane's colony to England. Grenville had left Devon on May 2, 1586, and, going via Madeira, reached Virginia in the middle of July. At Roanoke he found only Lane's deserted fort and houses, and lest England's claims lapse for want of occupation, he landed a small force with enough supplies to last a year. After some profitable privateering in Newfoundland and the Azores, he was back in England by Christmas. (Drake had reached Plymouth in July.) Of those he left behind at Roanoke the record holds the names of only the leaders (Coffin and Chapman, from Barnstaple in the West Country). Neither they nor their sixteen fellows were ever heard from again.

It must have seemed discouraging. As if in token, Lane wrote to Raleigh how, as Drake's boats were taking them off from Roanoke, the "weather was so boysterous, and pinnaces so often on ground, that most of all wee had, with all our Cardes, Bookes, and writings were by the Saylers cast over boord, the greater number of the Fleete being much agrieved with their long and daungerous abode in the miserable road." Thus were lost most of John White's drawings and Thomas Hariot's observations. Yet Sir Walter Raleigh, at the pinnacle of the queen's favor but now with more than ever on the line, would not relent. His propagandists (White, Hariot and Richard Hakluyt) set about trumpeting the word that Lane's experience and the information about Virginia now at hand proved beyond a doubt that it was a place fit for English habitation.

The composition of Raleigh's third expedition reflected both that conviction and a certain urgency finally to prove the point. With his own resources sorely stretched, Raleigh this time assembled a distinguished organization of merchants and gentlemen to back the effort, and he abandoned the notion of a merely military garrison run by soldiers like Lane. Instead he appointed artist John White as governor, and he was to rule with nine assistants, all colonists themselves. Under license from Raleigh, the "Cittie of Ralegh in Virginia" was duly incorporated on January 7, 1587. Sir Walter even got the Garter King of Arms to issue the colony a coat of arms (a red cross on a white field with Raleigh's own roebuck crest in the upper left quadrant), along with separate ensigns of honor for White and each of his assistants. White further thought to dignify himself with a new suit of armor. Would-be squires all, these colonists all came from plain middle-class origins in England and Ireland, and this time there were among the 115 who sailed fourteen families, accounting for seventeen women and eleven children. Two of the women, Margery Harvey and Eleanor

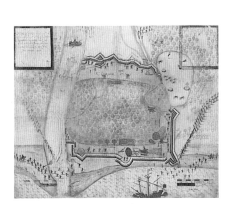

Artist John White, who accompanied Grenville's crew, drew the English colonists' fortified encampment in Puerto Rico, carefully recording some of their activities. (British Museum)

White Dare (the daughter of John White and wife of Ananias Dare, one of the assistants), were pregnant. Several of the men in addition to White were veterans of Lane's colony of 1585.

They sailed in three ships on May 8, 1587, which was late to make the Atlantic crossing and still plant crops for the fall. Their arrival in Virginia was further delayed by their pilot, again Simon Ferdinando, who resolutely preferred privateering to ferrying colonists and dallied through the West Indies looking for prizes. They made their continental landfall somewhere between Cape Fear and Cape Lookout, worked their way up the Banks and did not anchor off Roanoke Island until July 22. But it was the Chesapeake, not Roanoke at all, that was to be their destination. That they never got there was thanks to Ferdinando, who promptly put his passengers ashore at the site of the Lane colony, insisting that the season was too far gone to go hunting for something better farther north. (What worried him more was that he might miss his chance to go hunting for the Spanish plate fleet known to head out across the Atlantic in August or early September in search of gold and silver.) White was not the leader to stand up to his bullying, and even after it was discovered that the party Grenville left behind had been butchered by the Indians, White could get the Portuguese to go no farther. It was not a promising start.

The old fort had been destroyed, so the new colonists set about refurbishing Lane's houses, "overgrowen with Melons of divers sorts, and Deere within them feeding," and making the best of it. The first casualty came quickly, on July 28, when George Howe, one of the assistants, was fishing for crab not far from the village and was slain by Indians who "shotte at him in the water, where they gave him sixteene wounds with their arrows: and after they had slain him with their woodden swordes, beat his head in peeces, and fled over the water to the maine." Determined not to repeat Lane's mistakes, White forbore quick reprisal and tried to negotiate better relations with his hosts, but when he did take action at last it turned out to be against the wrong Indians. Despite faithful Manteo's best efforts, such trust as there ever was between the two races then was gone for good. But summer was still high, and spirits with it. Following Raleigh's wishes, Manteo was christened and White bestowed on him the title "Lord of Roanoke and of Dasemunkepeuc in reward of his faithful service." Five days later, on August 18, Eleanor Dare was delivered of a daughter, "the same christened there the Sunday following and because this childe was the first Christian borne in Virginia, she was named 'Virginia.'" Margery Harvey's child was born a few days later; we do not know whether it was a boy or a girl.

Stout hearts were one thing, full bellies another, and it was soon apparent given the approach of autumn and the unhelpfulness of the Indians that the supply situation was critical. The colonists therefore decided to petition Raleigh immediately for relief. All wanted to stay and make a go of it, and White himself, the only one of them thought to have any influence back in England, at

John White's map of Raleigh's Virginia rendered the coastline from the mouth of Chesapeake Bay to Cape Lookout with then-unprecedented accuracy. Decoration is limited to English ships and Indian canoes. (British Museum)

last agreed to return and serve as factor for the colony. On August 27, he and Ferdinando sailed for home in two of the three ships. Should they have to move from Roanoke in his absence, the colonists made arrangements that specific signs be left behind to point the way to where they had gone.

Raleigh was no more inattentive to White's report of the situation at his Cittie of Ralegh than he had been on Grenville's first return after planting the 1585 colony. He learned that supplies were low, the Indians hostile, and the Chesapeake still unattained, and by the end of March 1588 a relief fleet of seven or eight ships, commanded by Grenville, was ready to sail. Ordered instead by the Privy Council to join the Royal Navy and prepare to defend England against Philip's Armada, it never left the harbor. Raleigh did manage permission for a small bark and pinnace, the *Brave* and the *Roe*, to reinforce the settlement, but they were set upon by French pirates while still in European waters, and after a bloody fight that left White wounded, were forced to return to England. The Armada soon appeared off Plymouth, and all England's energies and all her sailors concentrated to meet it. Raleigh joined the fleet in the Narrow Seas, and Roanoke was left to fend for itself across the wide ocean reaches.

When White at last returned to Roanoke, in August of 1590, it was as a passenger on a privateer that, after hunting Spanish cargoes in the Caribbean, made a call in Virginia. He had left his daughter and baby granddaughter on this far shore, and his report of the scene as he once again (so he thought) approached them resonates with hope held out to the very last and then snuffed out: "We espied toward the North end of the Iland ye light of a great fire thorow the woods, to the which we presently rowed: when wee came right over against it, we let fall our Grapnel neere the shore and sounded with a trumpet a Call, & afterwardes many familiar English tunes of Songs, and called to them friendly; but we had no answere . . ."

At daybreak they landed and found evidence only of a forest fire and of where the colonists once had been: "We found the houses taken downe, and the place very strongly enclosed with a high palisado of great trees, with cortynes and flankers very fort-like, and one of the chiefe trees or postes at the right side of the entrance had the barke taken off, and 5 foote from the ground in fayre Capitall letters was graven CROATOAN without any crosse or signe of distresse; this done we entred into the palisado, where we found many barres of Iron, two pigges of Lead, foure yron fowlers, Iron sackershotte, and much like heavie things, throwen here and there, almost overgrown with grasse weedes. From thence wee went along by the water side, towards the poynt of the Creeke to see if we could find any of their botes or Pinisse, but we could perceive no signe of them, nor any of the last Flakons and small Ordinance which were left with them, at my departure from them. At our return from the Creeke, some of our Saylers meeting us, tolde us that they had found where divers chests had bene hidden, and

Raleigh, the last and most flamboyant of the Elizabethans, went to his death with characteristic dignity and grace. The lower half of this engraving shows his execution. (British Museum)

long sithence digged up againe and broken up, and much of the goods in them spoyled and scattered about, but nothing left, of such things as the Savages knew any use of, undefaced. Presently Captaine Cooke and I went to the place, which was in the ende of an old trench, made two yeeres past by Captain Amadas: where wee fulund five Chests, that had been carefully hidden of the Planters, and of the same chests three were my owne, and about the place many of my things spoyled and broken, and my bookes torne from the covers, the frames of some of my pictures and Mappes rotten and spoyled with rayne, and my armour almost eaten through with rust."

Sic transit gloria mundi, Garter coats of arms and all, White might have thought. Again a storm intervened, and he was unable to continue the search at Croatoan Island where Manteo's people lived and where it seemed the colonists most likely had gone. Badly battered, the ships returned to England, and that was the end of it. White retired to Raleigh's estates in Ireland and himself soon vanished from history's record. Raleigh's own fortunes soon tumbled. Elizabeth grew old.

As to the colonists, authorities speculate that some of them may have migrated north to where they had wanted to be in the first place, to the vicinity of Lynnhaven Bay near the present-day city of Virginia Beach, where they lived with friendly Indians for more than twenty years, only to be slaughtered by Chief Powhatan's warriors on the very eve of the Jamestown settlement in 1607. By then, Virginia Dare could have had children of her own, no doubt by an Indian father. Others of them likely did move south, as the carving "CROATOAN" indicated, and when help from home never came, slowly intermixed with Manteo's people and so too vanished. No other English returned to settle at Roanoke for a hundred years. There the facts of Roanoke's history cease: beyond John White's account of the 1587 voyage that planted the colony of men, women and children, we know for sure nothing else, either of what the colonists did while at Roanoke after White left, or of what became of them after they left, as they probably did.

Sir Walter Raleigh, the Roanoke colonists' patron, is one of history's finer characters. The pages of biography (there are several good ones) will not hold him. Paintings of him are striking in a way that suggests the man was more so. His involvement with Roanoke and Virginia was but one of many enthusiasms in a long, exciting life whose chapters reveal astounding range: plain yeoman origins in the West Country; self-taught to the point of great learning; soldier of fortune in France and Ireland; courtier extraordinary; "Darling of England's Cleopatra"; colonizer of the New World; dreamer after El Dorado; hero of Cadiz; historian of the world; poet of love and loss. After his queen, he was England's best defender, Spain's worst enemy; he was a scientist and patron of scientists; maker of plots and victim of plotters; ardent lover; mature husband; wronged prisoner in the Tower; martyr on the block; Church of England Christian.

A portrait of James I by Jan de Critz. (National Maritime Museum, Greenwich)

He has been called the last Elizabethan. He basked as did few others in the rays of Gloriana, and not even his secret marriage to Elizabeth's lady-in-waiting Bess Throckmorton, which so severely displeased the queen, cost him her friendship. When Elizabeth died three years into the new century, Sir Walter lived on but did not otherwise make the transition.

James VI of Scotland, who became James I of England, was emphatically not one of history's finer characters. He proved a wretched monarch, his misguided policies and un-English notions about the divine right of kings forfeiting him the trust of his new English subjects and making a dangerous enemy of their Parliament. Pro-Catholic in his religion and pro-Spanish in his foreign policy, James stood temperamentally and intellectually far removed from everything Sir Walter Raleigh had lived and fought for. In 1603 (the very year of James's accession) Raleigh was unjustly tried for treason for, of all things, conspiring with Spain, and was condemned. In his queer way, James then toyed with the great man for the rest of Raleigh's life, reprieving him from the block at the last moment but keeping him his prisoner, still under sentence of death, for the next twelve years.

After an early suicide attempt, Raleigh recovered his balance and put his time in the Tower to good use. He set up a small household within the walls, conducted elaborate scientific experiments, befriended James's eldest son, Henry, Prince of Wales (who, unfortunately for Sir Walter, died in his teens and long before his father), and wrote his remarkable *History of the World*, an elaborate lesson in the uses of the past for the instruction of the present. It went through ten editions after Raleigh's death and supplied historical context for England's anti-authoritarian revolt against the Stuart monarchy in the 1630s and 1640s. Oliver Cromwell, John Locke and thousands of more ordinary Englishmen took to heart its lesson that God acted in the political affairs of men and that he did intervene to punish unjust rulers. The king, though he never did mend his ways, also took its meaning and to the extent he could had the book banned.

Fear and loathe Raleigh as he might, perfidious James had one great need that it seemed, fantastically, his prisoner in the Tower might help him meet: money. He knew of Raleigh's wild adventure up the Orinoco River in Guiana in 1595 in search of the legendary golden city of El Dorado, and that Raleigh still claimed there was evidence that would lead to a fabulously rich mine there. Raleigh was as willing as ever to mislead to win his freedom (he in fact had no evidence at all), and James as craven as ever to replenish his treasury. So, in the twilight, the king released the old courtier in March 1616 with a commission (but no pardon) to raise a second expedition to Guiana, the loot to be split between the crown and Raleigh and his adventurers. So Sir Walter Raleigh set sail on his last earthly adventure. It turned into a catastrophe. A Spanish fort was mistakenly attacked; Raleigh's son Wat was killed; his longtime friend Lawrence Keymis committed suicide after Raleigh, in

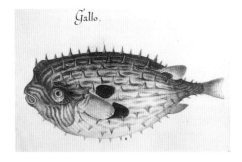

Burrfish by John White. (British Museum)

a fit of dejection, denounced his leadership. The sad episode finally sealed Raleigh's fate with James, then in the throes of negotiating a Spanish marriage for his second son and now heir, Charles, and he proceeded to sacrifice Raleigh to the demands of the Spanish ambassador.

On the scaffold, Raleigh relished his last moments and acquitted himself with the nobility of a great and faithful soul. He refused the blindfold and commented famously of the ax: "This is sharp medicine, but it is a cure for all diseases." It was said that the ghost of Sir Walter Raleigh stalked the Stuarts themselves to the scaffold; James died in bed, but his son Charles I lost his head to Cromwell's ax in 1649. As if in testimony to the relative worth of Raleigh and his Stuart adversaries, there is said to have risen from the crowd, as the bloody head was held high, the murmur: "We have not another such head to be cut off." Such was the popular judgment on the last Elizabethan, and England under James.

On the morning of his execution, after eating his breakfast, Raleigh took a pipe of tobacco. It was a fitting gesture from an avid smoker and promoter of Virginia's greatest product (and a private snub to James, who was a great enemy of the noxious weed). Raleigh lived to see Jamestown established, but we have no record of his reaction to its being named after his nemesis. He may, however, have taken pleasure in the fact that while in years to come the town of Jamestown returned to a forgotten backwater, the colony of Virginia, named for the lady who in his eyes could do no wrong, went on to great things.

Raleigh's own name ended up, somewhat less prestigiously, on the capital city of the state of North Carolina (spelled "Raleigh," although Sir Walter spelt his own name Ralegh) and on a pipe tobacco. On Roanoke Island today, there is no Cittie of Ralegh but only the town of Manteo, named after the Indian whom Amadas and Barlowe brought back to England in 1584 and whom Raleigh proudly paraded at Elizabeth's court. Yet the name Raleigh lives on in other ways, and still conjures all the fancied romance and adventure of the Elizabethan Age. His ill-fated colony that was produced by the Roanoke Voyages may indeed have been planted at the wrong place (where there was no good harbor) at the wrong time (when England's energies went to beating back the Spaniards), and when its communication with home was cut it suffered a sad but not surprising fate. The facts of that fate fail us, and because they do the speculation spun around them becomes so compelling. Interpretation of the colony's unfinished story as it is popularly rendered—the legend of the "Lost Colony"—is not fiction literally, for it does not rely on deliberate invention. It is based on the historical record such as it is and on "real" characters. But drama, not history, is its language, inspiration its purpose. On Roanoke Island for over fifty years now, Raleigh's lost colony has been a sizeable local industry and a study in the patriotic uses of a well-told tale.

History knows few men more patriotic than Sir Walter Raleigh.

Indians around a campfire, Indian woman and girl by John White. (British Museum)

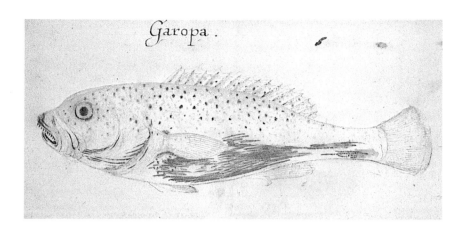

Grouper by John White. (British Museum)

He loved England and hated her enemies. I confronted him at Roanoke, as he has been presented by history's handlers, and found him enlisted into the service of another, much younger country. America is more self-consciously patriotic than most nations, so it is not surprising that Americans should have appropriated such a character as the mighty Raleigh, with his grand New World plans, for the purpose of fashioning a mighty myth accessible to all. I have heard it argued that the intensity with which men feel compelled to commemorate their past varies directly with their perception of the rate of change in their own culture: the more rapid and unsettling it seems to be, the more tempting the backward glance becomes. If so, then it is also so for the folk of Roanoke Island, whose backward glance has been deliberately fashioned, if not deliberately invented, into a celebration of specific events and a bold assertion of their relevance now.

It began in the late nineteenth century, when Roanoke Island was still an extremely remote and undeveloped place, and when local residents and a few summer people sought to publicize the fact that Raleigh's colony—and not Jamestown or Plymouth—was "first." But it was not just the historical fact that Raleigh's colonists were "first" that caused these rural Carolinians to take a band of obscure Elizabethans to their hearts; it was the romantic fact that they were "lost." They celebrated Virginia Dare's birthday as something of a local holiday, and under the direction of a local teacher, Mabel Evans Jones, mounted an annual "pageant-drama" around the story of Eleanor Dare and her baby daughter, John White, Manteo and Raleigh. Bigger things followed with a pageant composed for the 300th anniversary of Raleigh's execution in 1618. Written by Frederick Koch, *Ralegh, Shepherd of the Ocean* was presented in October 1920 during state fair week in Raleigh, North Carolina. As Koch conceived him, Raleigh represented "the struggle of the English people for freedom from tyrant rule, as blazing the way for those who came after him to inherit the fruition of his vision of a brave New World—the proved reality of his dream of a new 'English nation' in America."

At the same time, the North Carolina state department of education set about, with the help of Mabel Jones, making a silent film about the Roanoke Voyages. The five-reel production carried sixty-three captions narrating the story depicted on the screen, which began with Raleigh and Richard Grenville listening to an old salt tell tales of exotic worlds across the sea, and continued through Amadas, Barlowe, Lane, White and the colonists who disappeared into the wilderness. In 1925 Jones scripted a new drama, drawn from the movie, and produced it with local talent on an outdoor stage at the north end of the island overlooking Roanoke Sound. In 1933 she completed another version, titled *America Dawning.*

The production *The Lost Colony*, which you can see today during the summer season, stems from efforts to commemorate the 350th anniversary of the first Roanoke voyage in 1584 and the birth of Virginia Dare three years later. Through largely local enterprise, the Roanoke Colony Memorial Association commissioned prominent dramatist and North Carolina native Paul Green, who had won a Pulitzer Prize for his play *In Abraham's Bosom*, to write a new pageant-drama on the Roanoke story. Green's fee was $1,500, and the result—*The Lost Colony*—is a moving populist epic correct in its general historical outline and richly redolent of the New Deal atmosphere that surrounded it.

Paul Green went on to write other "symphonic dramas," as he called them, but *The Lost Colony* was where he proved the worth and the enduring popularity of the form. As he put it: "One for all and all for one, a true democracy" of poetry, dance, pantomime, choreography, story line and music. When I announced my intention to visit Roanoke Island especially to see the play, friends derided it as just another attraction served up for vacationers looking for a night's entertainment fit for the kids. That is indeed who comes: "we the people" casually arrayed for seaside fun. And come they do, to sit on hard wooden benches for nearly three hours of costume drama under the stars. There is nothing at all casual about the thoroughly professional production they see. The personnel has changed many times since the opening, but the script the company uses today differs very little from the one Green wrote back in 1937. The scene list recalls the plot. Act I: Prologue; An Indian Village on Roanoke Island, Summer 1584; A Tavern Yard in London, Some Months Later; England, Queen Elizabeth's Garden, The Same Day; King Wingina's Village on Roanoke Island, Summer 1585; England, A Wharf in Plymouth, Spring 1587. Act II: The Fort in the Cittie of Ralegh on Roanoke Island, July 1587; The Same, Three Weeks Later; The Same, The Following Sunday; The Same, Christmas 1587; England, The Queen's Chamber, Spring 1588; The Fort on Roanoke Island, Christmas 1588. The music sets a mood: old English folk songs, ballads and Anglican hymns rendered by the Lost Colony Choir. The characters tell the tale: Indian youths and warriors, maidens and milkmaid dancers, heralds, courtiers, pages, ladies-in-waiting, sailors, colonists, Raleigh and Elizabeth i. Even with its message set aside, it is all superb spectacle in a very beautiful place. But the message is overwhelming—and utterly faithful to Green's

Flying fish by John White. (British Museum)

intentions and to the wholesome patriotism of the early Roanoke memorializers. The old posturing trouper Raleigh would have loved it: once again, and again and again, he takes his world by storm.

A historian himself, Raleigh would have approved especially of the dramatic device of a narrator, who like Thornton Wilder's famous Stage Manager in *Our Town*, ties the pieces of the tale together. In *The Lost Colony*, the Historian establishes at the beginning the serious, reverential tone of the play: "For here once walked the men of dreams / The sons of hope and pain and wonder / Upon their foreheads truth's bright diadem, / The light of the sun in their countenance, / And their lips singing a new song— / A song for ages yet unborn, / For us the children that came after them— / 'O new and mighty world to be!' / They sang, / 'O land majestic, free, unbounded' . . . Now down the trackless hollow years / That swallowed them but not their song / We send response— / 'O lusty singer, dreamer, pioneer, / Lord of the wilderness, the unafraid, / Tamer of darkness, fire and flood, / Of the soaring spirit winged aloft / On the plumes of agony and death— / Hear us, O hear! / It lives, it lives, / And shall not die!" And at the end, as the remnant of the colonists march off into oblivion, he repeats: "Now down the trackless hollow years / That swallowed them but not their song / We send response . . .," etc., etc., etc.

Whatever exactly "the dream" was (for Green it was that "a nation of liberty and free men shall continue on the earth"), *The Lost Colony* presents it as almost a religious experience, somehow transcendent of ordinary history. Part of it, of course, Green had to make up (the part of the story that takes place after John White returned to England and before he got back to Roanoke in 1590), and it does have at least the transcendence of art. But attending a performance of *The Lost Colony* is like nothing so much as being in a great patriotic church, among a solemn assembly of the already evangelized, though it is much more beautiful. Green was a master dramatizer, as well as evangelist (as was Raleigh), who knew just how to set up an audience to receive his message and then send them home remembering it. Everyone loves a song, and *The Lost Colony* is filled with music, played on a great organ and sung in chorus. Ever eager for verisimilitude, Green went to period sources: Duncan's *Minstrelsy of England*, Chappell's *Old English Popular Music*, the *Oxford Book of Carols*, the *Episcopal Hymnal*. And he came away with fine old tunes: "The Battle of Agincourt," "Hast Thou Heard What Wise Men Say," "O Farewell England, Farewell All," "O Once I Was Courted," "The Holly and the Ivy."

Green placed a hymn, the same hymn—"O God That Madest Earth and Sky"—first and last. It is a curious selection. I do not know how he intended it, and can only testify as to its effect. No doubt he liked the poetry and found its cadence and imagery exquisitely "of the period." It is a dark hymn, darker by far than the inspirational words of much of the script, which is all about lusty singers, dreamers and pioneers.

O God that madest earth and sky
And hedged the restless seas around,
Who that vast firmament on high
With golden banded stars hath bound—

O thou whose mighty arm doth keep
The trembling world, the failing sun,
Whose shining presence fills the deep
Where lightless time's dark measures run—

O God our Father, Lord above,
O bright immortal, holy one,
Secure within thy boundless love
We walk this way of death alone.

The sun sets on Chesapeake Bay, the intended site of the second English colony in Virginia, chosen because it would provide arable land and a deep-water naval base for attacks on Spanish ships. However, a fateful change in destination took the colonists instead to Roanoke Island. (Steve Dunwell/The Image Bank)

The sentiments are wholly Christian, and like Raleigh on the scaffold, void of skepticism and spiritual unease. They belong to religion in an age before religion became confused with politics, ideology, patriotism, as it was even in Green's day. There is nothing modern about these sentiments.

Nor is there much modern about Raleigh, whatever the commemorators make out of his oft-quoted line about living to see a "new English nation" planted in America. We do not know if Raleigh ever sang "O God That Madest Earth and Sky," but we do know that his life had led him, if not to Virginia, then to the same serene faith and that that faith perfectly consoled him at the end. In the Abbey Gatehouse at Westminster during the final hours of his life, he inscribed on the flyleaf of his Bible a last melancholic stanza to an otherwise passionate poem he had written to his beloved Bess Throckmorton in the late 1580s, and so composed his and, if you like, the Roanoke colonists' epitaph:

> *Even such is Time, which takes in trust*
> *Our youth, our joys, and all we have,*
> *And pays us but with age and dust;*
> *Who in the dark and silent grave,*
> *When we have wandered all our ways,*
> *Shuts up the story of our days:*
> *And from which earth, and grave, and dust,*
> *The Lord shall raise me up, I trust.*

If you climb to the top of Jockeys Ridge near Nags Head on the Outer Banks, you can look east over the "restless seas" that reach to England, and west over the shallows of Roanoke Sound to the lost colony. It was across this spit of sand that the Roanoke voyagers portaged all their little bits of England, from little ships anchored in the open ocean, across to the Cittie of Ralegh in the wilderness. In the words of the hymn, "death alone" awaited them. We do not know whether they met it as nobly as their great patron later met his. Paul Green's *The Lost Colony* offers one answer. I am satisfied to think that they did well enough against bad odds, and that the world they discovered on this low and sandy shore simply was not ready to accommodate them. Pioneers? Or just "mere English" cast away far from home?

The first English to make a home away from home in America, the 1587 voyagers died trying, though perhaps not any sooner than many would have died in England anyway. Discovery cost thousands of lives (and it cost the Indians their very way of life). Because of where and when they died, history has dealt kindly with these few score Elizabethans and, where history is silent, imagination has beneficently filled in the gaps. But even in the most dramatic rendering, with costumes, lights, and set to music on the very place where it all likely happened, their story still strikes one not as tragic, only lonely.

If you travel a bit farther down the Banks, to Ocracoke Island, you will find in Ocracoke village evidence of other Englishmen who left their bones on this lonely coast: four sailors washed ashore from the HMS *Bedfordshire*, which was torpedoed by a German

submarine here on May 14, 1942. Two are known (Lieutenant Thomas Cunningham, twenty-seven, and Ordinary Telegrapher Stanley Craig, twenty-four), two unknown. Their remains lie in the Howard-Wahab Graveyard and are marked by four white crosses in a picket enclosure. The Union Jack flies overhead. On a small brass plate Rupert Brooke's quiet lines, known by every true-blue Englishman in the generation after World War I, supply their epitaph:

The fort at Roanoke Island, in what is today eastern North Carolina, was built by the English in the late 1580s as protection against both Indian and Spanish intrusion. But when communications with the mother country failed at a crucial moment, no fortress availed the beleaguered English far from home, and the Roanoke colony slipped from history and entered legend. (U.S. National Park Service)

> *If I should die, think only this of me:*
> *That there's some corner of a foreign field*
> *That is for ever England.*

It seems an epitaph also for the Roanoke colonists, at least as fitting as *The Lost Colony*'s lofty lines about pioneers in a land "majestic, free, unbounded" with a "song for ages yet unborn." To this day, the plot is leased by the State of North Carolina to the British Government and kept tidy by American sailors of the United States Coast Guard, Ocracoke Station. Nearby, the oldest lighthouse in operation in America (built in 1823) still shines its warning to sailors far from home.

Jamestown Ferry

FROM THE HIGHLANDS AND FOOTHILLS OF THE
Appalachian Mountains five rivers flow down to tidewater at the
Chesapeake Bay. None reaches very far inland, and the Bay itself
is just that, a bay and a dead end. No passage to India here for
spice-hungry Europeans, though for a time it must have looked
pretty promising: the Chesapeake is one of the greatest sheltered
inlets of the ocean anywhere in the world. Tucked in at its foot,
between modern-day Norfolk and Newport News, also lies one
of the world's greatest natural harbors.

A good harbor, in the sail-borne age, was a much-coveted thing.
And the Atlantic coastline of North America in this respect was
a welcoming one for the venturing mariners of the sixteenth and
seventeenth centuries. You can, if you poke around enough, get
in off the sea. Four of those rivers, it is fair to say, belong emo-
tionally and geographically to Virginia. (North to south: the
Potomac, the Rappahannock, the York and the James; the fifth,
the Susquahanna, is very much a Pennsylvania affair.) Together
they make Virginia's shore, once you find your way to it through
the opening between Cape Henry and Cape Charles, especially
hospitable, and together they have helped give Virginia a special
place in the history of discovery.

For Virginia became site of the first English settlement in North
America that stuck, that, once planted, did not vanish or have to
be withdrawn. I learned about this when as a Virginia schoolboy
in the 1950s I received what seemed at the time rather more than
the usual lackluster instruction in state history. Virginia then took,
and may still take, its past seriously. By North American standards, there is a lot of it (close to four centuries), and it was obviously deemed by the people who decided such things to hold for
the youth of the day many worthy lessons and inspiring exam-
ples. Most of them, to my discredit, I can no longer recall. One

A replica of the Susan Constant *lies
docked at Jamestown. (Russ Kinne/Miller
Comstock)*

that I do is Jamestown. When I first saw the place I was nine, and it was 350 and celebrating with the opening of the Festival Park that still welcomes school-age and grown-up visitors today. Then, I came to it on a bus with my classmates. This time, I approached alone and, as the English first did, by water.

The James River where it reaches the Chesapeake Bay is five miles across and deep enough to accommodate the commerce of the world. Where Englishmen once passed upstream, today chiefly coal passes down, from the deep mines of the western counties of Virginia and West Virginia, brought on long trains that roll right up onto the docks to be emptied like toys into the gaping holds of bulk carriers from such resource-poor places as the Philippines. Above the coal docks for nearly a mile along the north bank reach the mammoth cranes and building yards of the Newport News Shipbuilding and Drydock Company, down whose ways have slid the merchantmen and men-of-war that for years have carried the colors of this New World nation to the ports and coastlines of many old ones. Nuclear submarines, today's dreadnoughts and ships-of-the-line, are built and refueled here, their sinister whale-like hulls hardly visible along the cluttered shore. Just upstream, the old James River Bridge joins Newport News with Isle of Wight County and announces to the eye where ocean harbor ends and inland river begins. The river bends past Fort Eustis and around protruding Hog Island Game Refuge, and bears west toward the low-lying peninsula where Captain Christopher Newport led his little fleet in May 1607.

The State (or Commonwealth, as they officially say) of Virginia operates a ferry service across the river from Scotland in Surry County to Jamestown. Carrying cars, campers, bikers and even a few foot passengers (there is no bridge for the nearly fifty miles that separate Newport News and Hopewell), it makes a gentle S-curve into the current, passes Jamestown Island on the starboard side, and lands in twenty minutes or so near the Festival Park, where State Route 31 runs down to the river's edge.

Though there has been a ferry here since 1925, the boat I rode was a new one, with all the modern rules. You can't smoke (except in your vehicle) and Consumption of Alcoholic Beverages is Prohibited. But they do let you look, which is something that passage on any boat—even the humblest shortest-hop ferry—seems universally to encourage. And it was from the rail of the MV *Williamsburg* that I saw again "the three ships," Virginia schoolboy shorthand for *Susan Constant, Godspeed* and *Discovery*—only replicas, for sure, of Captain Newport's seventeenth-century fleet but, seen across the calm James water under a lowering evening sky, very credible ghosts indeed.

Stout little tubs built for the rough and tumble of trade, which was the business that Jamestown from the start was meant by its owners to be part of. The origins of this first permanent English settlement were pre-eminently commercial, though in the background played all the themes of politics and religion that had con-

FRANCISCVS DRAECK · NOBILISSIMVS
EQVES ANGLIAE · IS EST QVI TOTO T
TERRARVM   ORBE CRCMDVGO

id circumdūcto pernosco
in fongitudine, in fatitudi·
ne est Jmpossibile, etc:

Dressed in armor and grasping a musket, Sir Francis Drake is depicted as "England's most renowned knight," according to the Latin inscription. (John Carter Brown Library, Brown University)

torted the history of Europe since the Reformation. It was as a business proposition that colonization first captured the imagination of the English, and without commercial success, not all the power of patriotism and religion together could have sustained this, their first successful effort at Jamestown. Old England on a new shore—not a city on a hill—described what its founders had in mind for this colony. In time, their colony of Virginia spawned just that: a deferential social order with parishes, vestries, justices of the peace, sheriffs, an aristocratic legislature and a surrogate king. The denouement of decades of discovery, Jamestown was just how everything was supposed to turn out.

Jamestown, as depicted in a contemporary painting, took several decades and much trial and error before proving that permanence and prosperity were possible for Europeans on the North American shore. (Colonial National Historical Park)

The men responsible for it and in whose image it was cast were fired by the considerable corporate energy of the late Elizabethan Age and their own healthy desire for gain. Jamestown stood, if you will, at the early dawn of a new imperial age, which lasted for the next three centuries and which for the glory of God, the welfare of England and their own individual advancement, sent thousands of English to tame the far corners of the earth.

In 1606 one group of them from the West Country—Sir Francis Drake's England—petitioned the new king, James I, for license to colonize the New World. With Sir Walter Raleigh's exclusive patent from the 1580s now lapsed, the legal path was clear, and James, still advised by Elizabeth's great counselor Robert Cecil, assented. Most of the adventurers were seasoned veterans of Elizabeth's long wars with Spain: Sir Thomas Gates and Sir George Somers, Devon and Dorset men; Edward Maria Wingfield, a Catholic whose middle name kept alive memory of Elizabeth's sister, Mary Tudor; Ralegh Gilbert, son of Sir Humphrey Gilbert, who in 1585 had claimed Newfoundland as an English colony but could not settle it; George Popham, relation of Sir John Popham, Puritan Lord Chief Justice of England. The patent as granted by James envisioned two settlement enterprises. The northern, which was to lie between 38 and 45 degrees north latitude, in time received most of its backing from the trading and fishing communities of the West Country, chiefly Bristol, Exeter and Plymouth, and would be associated with the settlement of

New England, still then thought of as the northern part of Virginia. The southern, which concerns us here, was backed chiefly in London and was to be planted between 34 and 41 degrees north, roughly between present-day Wilmington, North Carolina and New York City.

"Backing" meant just that: financial support and management. Though the state weighed in with a clear crown warrant meant to warn the rest of the world that this was an English national enterprise and there was to be no trespassing, neither king nor Commons took the big risk and footed the bill. Virginia was to be a public joint-stock venture, which meant in practice that the heavy financing (approximately £200,000 between 1606 and 1624, when Virginia became a royal colony and the original joint-stock companies were dissolved), came from City of London merchant magnates such as Sir Thomas Smythe of the British East India Company. This was a notably different structure of support from the more personal, slapdash, gentry-backed efforts that had been behind the ill-fated colony at Roanoke Island twenty years earlier, and it proved flexible enough to see the job through despite an array of early disasters. To make money out of settlement with recognizably capitalist techniques was the hope that founded Virginia, and it proved a very durable hope at that.

But why just then? In my reading on this subject it was refreshing to learn from one of its major authorities, David Beers Quinn, what would seem not at all surprising and indeed almost common sense. That is, that a lot must remain speculation: too much passed between the principal characters, now nearly four hundred years ago, that was never written down. What we do know about the motives of the individuals involved we must try to infer from the general tenor and condition of the age.

The end of the sixteenth century brought to England both relief and trepidation, perfect springboards for serious colonization. The English owed relief to their defeat of the Spanish Armada in 1588, which secured their country's independence as a Protestant nation-state. After years of on-and-off hot-and-cold conflict with the dons, Elizabeth's great captains led by Drake "drummed them up the Channel" and won an epic national triumph. Without victory over the Armada, the confident spirit and concentrated effort that made Jamestown a success would have been hard to summon. Peace came only slowly and fear of the Spaniards lingered, but the balance had clearly shifted in England's favor.

One of the issues that Elizabeth had fought for was an open door to the Americas. And while Spain would still hold the South American continent until the national revolutions of the nineteenth century, her claims of exclusivity over lands not in effective occupation north of Florida, where there was as yet no solid settlement of any European power, now grew hollow. At the negotiations that ended the war, nothing whatever was said about English colonization in America, which suited the English, who soon went right ahead and did as they pleased. From the correspondence of

Discovery *was the smallest of Captain Christopher Newport's three-ship fleet that transported the colonists to Jamestown. (Jamestown-Yorktown Foundation)*

the Spanish ambassador in London, we do know that consideration was given to quashing Jamestown by force in its fledgling stages, a move that probably would have provoked another war. The Spanish demurred—and the opportunity never came again.

The thrill of military victory such as followed Armada Year came side by side with the other Elizabethan achievement: peace and security at home even as one reign at last gave way to another. In the age of dynastic politics and before the development of the idea of a loyal opposition, questions of succession could provoke bloody intrigue if not civil war. But when the childless Gloriana passed away in 1603 and was succeeded peacefully by her Scottish kinsman James Stuart, an added air of self-congratulation must have suffused the ranks of the English nation. No one then could know how much James would disappoint and that the next decades would see England split into two religious camps and go to war with itself. For the moment, however, the signs were propitious for bold ventures.

For the London merchants like Smythe, chief among those ventures was Virginia, which had been neglected since Raleigh's ill-fated Roanoke ran afoul of court politics and the distraction of the Armada. As a principal of the British East India Company whose first fleet returned richly laden to England in 1603, Smythe inspired the business confidence that led his City colleagues to invest. The king's chief minister, Lord Salisbury, and Lord Chief Justice Popham were crucial. Popham was friend of Ferdinando Gorges, who commanded the Plymouth garrison and with his grandson Thomas Hanham had important ties to Bristol.

The highly speculative nature of the enterprise and the speed with which it attracted investors point to the new attitude toward capital and investment that accompanied James's accession. Out from under the war with Spain, the City gamblers rolled the dice and rolled up their sleeves. Speculation—risk-taking in the later capitalist mode—looked for new objects of desire and, in the worlds laid open by a century of discovery, found them. The British East India Company had made a profit just three years after its founding in 1600. As peace returned there followed a new boom in the Newfoundland cod fisheries, where Europeans had been letting down their nets for a hundred years at least. We know that fully a third of the members of Parliament in the first seven years of James's reign put their own money into speculative ventures as, we would say today, "passive investors": eager as ever for good return but content with no direct involvement in the operations they backed. As landowners at home, many were also drawn to land speculation in the New World (as they were in Ireland), with or without the prospect of their own resettlement attached. Although England at the beginning of the seventeenth century could hardly be described as overpopulated, the profits in land that might be "developed" far from home naturally attracted the gentry of a small country where land could only grow more scarce. Of all the kinds of economic activity (save one) that attracted men to Virginia, land speculation cast the longest shadow, years after Jamestown itself had reverted to a backwater ruin.

That other activity was the cultivation of tobacco, an economic panacea that in the first years was unexploited. From the first, however, the backers in London brought to the task at hand the merchant's unwavering understanding that the success of any colonial venture depended on the hard facts of economic performance. "New" Englands planted on far shores would have to justify themselves with economies that somehow complemented England's own. In this early stage of settlement, the attention of the Londoners was conditioned by the high profits to be had from trade in commodities—spices, wines, fruits, dyes—associated with Eastern, Mediterranean or Iberian origins. The peace with Spain had in this regard been disappointing. North African pirates still raised risks on the old routes to the Levant; Spain and Portugal remained backward and mercurial trading partners; and Spain's rich American possessions remained off-limits to anything but Spanish shipping. This was a prohibition more honored in the breach, and it spawned the golden age of Caribbean buccaneering; it was also a frustration with larger, longer-lasting consequences.

Why could not a colony somewhat to the north, say in Virginia, fulfill some of the same needs? The answer was that it could, if it found something to produce for markets the London merchants serviced. Much was tried—forest products for naval stores, barrel staves for the wine trade, cedar for cabinetmaking, sassafras for the pharmacopoeia, mulberry trees for silk—until tobacco emerged by the 1620s as the cash crop that the world wanted to consume and that Virginia wanted to produce. Whatever the source, whether in agriculture, industry or even mining, something had to be extracted from the New World that had value in the context of established economic activity back home. The London merchants were on the prowl for opportunity, and a permanent colony, at a place like Jamestown, answered their needs as no mere outpost could.

Their behavior toward Jamestown proved they were in it for the long run, and though they lost direct control of it in less than twenty years, their unfaltering lead gave the colony the time it needed to become a success. Early tales about Jamestown's resources waxed extreme in the tradition of much discovery reporting; some even held out the old elixir of gold. Not many believed that, at least not for long, but they did believe that enough was out there to justify the investment and effort necessary to create wealth from nature's crude resources.

Sir Thomas Roe, an investor whose words suggest a characteristic Elizabethan combination of level-headedness and largeness of patriotic vision, said that Virginia in 1607 seemed "a land ready to supply us with all the necessary commodities naturally wanting to us, in which alone we suffer the Spanish reputation and power to swell over us" and so it would surely bring "honor and profit to our nation."

Tobacco became the staple crop of the new colony of Virginia and enabled the settlers to survive economically. (British Museum)

The man who reported the gold where there was none was Christopher Newport, captain of the seaborne stage of the Jamestown expedition. A more reliable sailor than assayer, he was forty years old when the three ships slipped down the Thames five days before Christmas 1606 and was one of the most esteemed seamen of his day. Like many who sailed with him, he had had military experience in the wars with Spain, and in the next four years would make a total of five voyages to Virginia. His fleet, chartered by the London Company, were merchantmen well built for the job of transportation ahead. With lengths only twice their beams, they were fatter and slower than men-of-war, but were also enormously sturdy and stable. Though this was a peacetime voyage, the two largest carried small artillery pieces suitable for use against pirates or Indian canoes.

At 120 tons burden, *Susan Constant* was the largest and carried in addition to fifty-four passengers and seventeen crew (including Newport), deep in her lower hold, English oats, barley and wheat for seed, all manner of tools and provisions, beer, wine and the makings of the meager fare that would feed everyone on the long outward voyage. *Godspeed*, first of her two small consorts, was forty tons burden, captained by Bartholomew Gosnold, with thirty-nine passengers and thirteen crew. Tiny *Discovery* under John Ratcliffe was a twenty-ton pinnace, built for coastal shipping in Europe and well suited to her new role of coastline exploration. The immense weight especially of *Susan Constant*'s timbers meant relatively deep draft—fully twelve feet, it was learned on reconstruction—which was one consideration in the siting of a settlement. The James River at Jamestown Island was deep enough inshore to permit mooring the ships to the branches of overhanging trees.

For their passage Newport chose a long and prudent southern path to avoid beating against prevailing Westerlies farther north and to break the voyage at known ports of call. George Percy, a younger son of the Duke of Northumberland, left the fullest account of the journey. It began with a slow start, as heavy storms forced them to heave to in the Downs still in sight of England: "The winds continued contrairie so long that we were forced to stay there some time, where we suffered great stormes, but by the skilfulness of the Captaine we suffered no great losse or danger." When the weather improved they coasted south off France and Portugal, *Godspeed* and *Discovery* putting in at the Canary Islands to top off their water casks. Just below the Tropic of Cancer, Newport signaled the fleet westward away from the African coast and out across the Atlantic.

Three months out from London, on March 23, 1607, they reached Martinique and then in quick succession paid calls at Dominica, Guadeloupe and Nevis in the Leeward Islands. On Guadeloupe they marveled at hot springs that cooked meat in half an hour; on Dominica they traded with natives (the usual beads and hatchets for potatoes, pineapples, bananas and, prophetically,

Indian man and woman eating by John White. (British Museum)

A romanticized vision of the landing of the Virginia settlers at Cape Henry in April 1607: the surf almost certainly would have been higher. (Colonial National Historical Park)

tobacco). On April 4 they sailed past what Percy called Ile of Virgines, probably St. John or St. Thomas, and stopped at Mona, between Puerto Rico and Hispaniola, where after exploring ashore passenger Edward Brooke became the expedition's first fatality: "His fat," as Percy put it, "melted within him by the great heat and drought of the Country." This too was prophetic, for other natural perils (though certainly not drought) soon claimed many of Brooke's fellows at Jamestown.

Yet for a short while longer, all must have seemed right with their world: blue seas, bright skies, fair wind and fresh provisions—and Virginia, the object of it all, now not far. One spring squall delayed them on April 21, but five days later they made their continental landfall at Cape Henry. Wrote Percy: "The six and twentieth day of April about foure a'clocke in the morning, wee descried the Land of Virginia; the same day wee entred the Bay of Chesupioc directly." Released from their long confinement, the colonists led by Reverend Robert Hunt promptly gave thanks for a safe passage. What they found of nature's gifts must have further lifted their spirits: "Faire meadowes and goodly tall trees, with such Fresh-waters running through the woods, As I was almost ravished at the first sight thereof." Man's first gifts were not so propitious, as inhospitable prior tenants emerged from the goodly tall trees and let loose with arrows that wounded Gabriel Archer and Mathew Morton.

Other hazards, of a political nature, soon became apparent. By the Company's orders, as long as the expedition was at sea Newport had sole supreme command. But once ashore at their destination, he was immediately to open the sealed box that contained the Company's identification of the "Councilors" who were jointly to govern Virginia in their stead. Henceforth (until he returned to England with *Susan Constant* and *Godspeed* in June), Newport shared authority with other Virginia adventurers: Captains Gosnold and Ratcliffe; Edward Maria Wingfield; John Martin, who had been a commander with Sir Francis Drake; George Kendall, whose cousin was Sir Edwin Sandys, who would be governor; and Captain John Smith, who was confined to the brig on charges brought by his fellows during the voyage.

It proved not to be a good system for putting backbone in a tiny colony on a far and hostile shore, and lack of forceful leadership very nearly lost Virginia its tenuous early lease on life. Ten years later, when the colony was still not out of the woods, planter John Rolfe bitterly bemoaned that in the beginning "this plantacion was governed by a President and Councell aristocraticallie. The President yerely chosen out of the Counsell, which consisted of twelve persons. This government lasted about two yeres: in which time such envie, dissentions and jarrs were daily sowen amongst them, that they choaked the seedes and blasted the fruits of all mens labors."

There was no precedent, and the owners back in London were feeling their way as much as the men from the three ships were feeling theirs. Less constitutional than organic, the problem partly lay with the character and composition of this particular group of settlers, which was heavily skewed with professional military men. Because the founders still took seriously the French and Spanish threats (the French had had a foothold at Port Royal in Nova Scotia since 1604 and were rumored to be probing southward, while the Spanish had been ensconced at St. Augustine in Florida since the 1560s and still nursed old grievances), it was natural that the first settlement contingents should incorporate experienced and disciplined soldiers. It has become one of the clichés of Virginia history that Jamestown's early population contained a surfeit of lazy gentlemen and ne'er-do-wells unfit for the hard work of settlement or for much else. To the degree that some of these "gentlemen" were highly professional men of arms and therefore poorly fitted to play other more immediately productive roles, the judgment is harsh. Good soldiers do not necessarily make good farmers, craftsmen or laborers, and these soldiers were no exception. Yet it was the quite soldierly virtues of one of them—Captain John Smith—that disciplined the colony back from the brink in its darkest hours and saved Virginia for the civilians.

But in April 1607 such conflicts were surface ripples only, as the leaders sent eager exploring parties to reconnoiter: Virginia Beach on April 27, Lynnhaven Inlet on April 28, Point Comfort on April 30. On April 29 they erected a cross at their landing place, the southernmost promontory on Chesapeake Bay, and named it Cape Henry after Henry, Prince of Wales, the oldest son of James I.

The place where these adventuring soldiers of the king waded ashore in 1607 is surrounded today, fittingly, by an army base: Fort Story, where the soldiers of the Republic go about their routine peacetime business amid old-fashioned white-clapboard, brown-trimmed, screened-porch-fronted buildings off the set of *From Here to Eternity*. Perched on sand dunes like the one that the Virginia adventurers hiked up to get a first hungry glimpse

The old Cape Henry Lighthouse at the southern tip of Chesapeake Bay in the present-day city of Virginia Beach, marks the site of the first mainland landing of the Virginia colonists in April 1607. (Pavilion Convention Center)

of their new world, Fort Story suggests an old-time anxiety for coastal defense, situated at the entrance to Chesapeake Bay and guarding the navy base beyond. Coming to it by road, I looked out the other way, across the bay and onto the ocean and spied no enemies real or remembered, no Russians, no Spaniards—only friendly warships returning from Atlantic patrol or tense tours in the Persian Gulf, and the toing and froing container ships of modern commerce flying many (and mostly) foreign flags. I paid my visit in early morning, when a hush hung over the place (it sits cheek-by-jowl with the bustling resort strip just to the south), something I would like to think is typical: this could not any longer be a very busy or important military bastion. Rather, its leftover 1930s feel makes it a good place for monuments.

Three bear on this subject. Reassuringly (for here land and sea meet dramatically) looms the old Cape Henry Lighthouse. Now disused but owned and maintained by the Association for the Preservation of Virginia Antiquities, it was the first lighthouse authorized by the United States Congress. Erected in 1791, it lit the entrance to Chesapeake Bay for ninety years. Unlike the more famous towers to the south along North Carolina's Outer Banks, this one still welcomes the hearty climber and opens a vista that Christopher Newport and company would have given much to behold: before it a vast expanse of bay and ocean; behind it, low vegetation, sheltered inlets and (today) the sprawling city of Virginia Beach.

So would have George Washington, the Marquis de Lafayette and General Lord Cornwallis, adversaries grimly faced off on the Yorktown Peninsula during the last act of the Revolutionary War. For from its light they could have seen their fates unfold before them in the Battle off The Capes, fought here between French and British squadrons on September 5, 1781. In a naval duel of classic form—battle line to battle line—the French Admiral Comte de Grasse forced the British to withdraw, which prevented reinforcement of Cornwallis at Yorktown and ensured an American victory. Near the lighthouse and the officers' club, the plaque that recounts all this was put up just in 1976, during the bicentennial of American independence, when a lot of historical commemorating got done.

The third monument is the Cape Henry Memorial Cross, erected in 1905 by the Daughters of the American Colonists to serve as a "reminder of that original oak cross planted by Englishmen in search of gold, adventure and natural resources. More importantly, Cape Henry marks the beginning in a continuing chain of events which saw the shaping of American culture by English institutions and customs." It is a candid, gently inspiring message, as honest as the clear view from the lighthouse. The original cross, long vanished, we are told may have been "of English oak fashioned for the purpose." If so, this would not have been unusual—many early discovery expeditions carried markers of one sort or another to stake their claims for king and country. It is only the inclusion of this sentiment on this marker that struck me, again,

The Memorial Cross at Jamestown, although it does not exactly mark the spot, evokes the piety of the seventeenth-century English who packed prayer books and bibles in their kit alongside the pikes and arquebuses. (Colonial National Historical Park)

A small-scale reconstruction of the Jamestown fort gives a sense of colonial life. (Russ Kinne/Miller Comstock)

as coming very close to the tone in which the men who raised their thanksgiving here, likely with cadences from Thomas Cranmer's 1549 Book of Common Prayer, would have wanted to be commemorated: as Englishmen devout and adventurous at once, who planted a bit of England far from home.

Duly landed and with God duly thanked, the Englishmen proceeded with the rest of their instructions from London: not to settle closer than thirty or forty miles from the coast and to plant coast-watchers to warn of unwanted intruders. They assembled a small shallop or rowboat for probing shoal waters, and for the next two weeks explored the James as far as its confluence with the Appomattox, near modern-day Hopewell where the bridge now is. On May 13 they settled on Jamestown Island, because it offered deep anchorage and because, as an island (or very nearly one), it was more easily defendable from land and sea attack. A low swampy place with only marginal sources of fresh water, it was not otherwise a good choice, as deaths from disease soon would show.

What happened next must have been attended by much excitement and relief at finally being able to get on with the job. Trees were cleared, the English seed grains planted (in time, they hoped, for a fall harvest) and shelter begun. The fort was completed by mid-June and bore the marks of its military builders: a triangular palisade enclosing about an acre, surrounded by a moat, with a bulwark at each corner mounting four or five demiculverins, small artillery that fired nine-pound balls. The main gate faced the river, not the land, and, as were the smaller openings, was protected by a piece of ordnance. Within the enclosure a church, storehouse, armory and communal living quarters were built of local materials and arranged along a fixed "street" and an open yard. Rough pine and oak made the frames, which were then filled with latticework of saplings and reeds. Wattle and daub, a crude plaster of mud and ground seashells, covered the walls. Reed roof thatching, as one would find at home in England, kept out the rain.

On the Company's instructions to seek out a water route through the American barrier to the Great South Sea (as the Pacific was then called) and the Orient, Newport and a small party probed upriver as far as the falls of the James, where Richmond would grow. In their absence, Indians attacked the settlement but were repelled by the ships' cannon. Soon a watch was mounted, helmeted halberdiers or pikemen stood guard, and at night the gates were tightly shut.

Fair weather prevailed, shoots of young wheat sprouted, the natives moderated their behavior, John Smith was released from confinement and as Captain Newport prepared to weigh anchor for England and resupply (leaving *Discovery* behind with the settlers), all seemed remarkably promising. Reverend Hunt celebrated the first Anglican communion in America in the open air on June 21, the third Sunday after Trinity. "Captain Newport dined ashore with our diet," wrote George Percy, "and invited many of us to supper as a farewell." The next day they were alone.

It was a deceptively smooth start, and the next few years of Jamestown's history were a lesson in just how tenuous their hold on the New World was. Slow, and culturally reluctant to learn the New World's ways, and not yet powerful enough to impose their own ways on it, these first English in America made for a portrait in precariousness. That summer the heat descended like a blanket, food spoiled, water went brackish, work discouraged the weak-willed, death and dissension struck. The first council president, Edward Maria Wingfield, proved no leader; his two quick successors, John Ratcliffe and Matthew Scrivener, were little better. Back in England, Newport may have talked loosely about gold, but reports from the colonists themselves, later compiled by William Simmonds, came closer to the truth: "Our drink was water, our lodging, castles in the air. With this lodging and diet, our extreme toil in bearing and planting palisades strained and bruised us. Our continued labor in the extremity of the heat had so weakened us as were cause sufficient to have made us miserable in our native country or any other place in the world. From May to September those that escaped dying lived upon sturgeon and sea crabs. Fifty in this time were buried." By the end of the first winter, 38 of the 104 who had come ashore were left alive.

The Company, however, was deadly serious about the venture and not easily discouraged. Between 1608 and 1610 it mounted three resupply missions bearing instructions, food, equipment, news from home—and new recruits, including even a few women. In return it called ever more insistently for profitable produce, which was scarce. The whole place burned to the ground in January 1608, planting was delayed by a short and futile gold rush, and were it not for the forceful and timely measures imposed by Captain John Smith, which have become part of the Jamestown legend ("If any would not work, neither should he eat"), the whole effort

Inside the reconstructed James Fort, modern-day Virginians have recreated a seventeenth-century English village in the wilderness. (Jamestown-Yorktown Foundation)

would likely not have survived the second winter. Threatening to hang anyone who tried to steal away by sea to Newfoundland, Smith initiated a fruitful trade with the Indians and generally disciplined the chaos. "By his own example, good words, fair promises," reported Simmonds, Smith "set some to mow, others to bind thatch, some to build houses, others to thatch them, always bearing the greatest task for his own share."

Smith was still in command at the time of the third resupply, which at once breathed new life into the colony but also nearly broke its slender back. The relief mission was the result of a great wave of popular enthusiasm back home for the Virginia enterprise. The council in London had swelled in membership to include Sir Oliver Cromwell, uncle of the future Protector; Sir Fulke Greville, friend of Philip Sidney; and Sir Maurice Berkeley. Investors high and low and a fair sampling of the wealth of the nation mushroomed: fifty-six companies from the City of London and 659 individuals, peers, knights, gentlemen, the patrons of Shakespeare, Cecils, Spencers, Chamberlains. There were Anglicans, Puritans and Catholics, and some settlers themselves who evidently preferred the aura of ownership to the stigma of hire.

But it was just then too much of a good thing. Supplies in the summer of 1609 were not great, but neither was the number of mouths to feed, and with Smith in charge they might have pulled through without catastrophe. That August, though, the resupply brought 400 more souls, green to the hardships of life in the wilderness, fever- and plague-ridden, their own supplies much damaged, their leaders separated from them by a hurricane and temporarily marooned in Bermuda (an episode from which sprang Shakespeare's *The Tempest*). Smith meanwhile was packed off to England by bickering opponents on the council, and the events of the ensuing winter soon entered the Jamestown canon along with the noble deeds of that poorly used soldier. "The Starving Time" it is called, when with hapless George Percy in charge, sickness, malnutrition and hostile Indians reduced the colony from 500 to 59. When the new governor, Sir Thomas Gates, at last arrived in May 1610 from his year-long diversion to Bermuda in two small and aptly named ships—*Patience* and *Deliverance*—what greeted him must have been heartbreaking. So much come to so little so quickly: Jamestown "raither as the ruins of some auntient fortification then that any people living might now inhabit it." Himself ill supplied, he opted for home and on June 7 embarked what was left of the settlers and floated down the James.

That might have been the end of it had it not been for the fortuitous (they would have said Providential) meeting with yet another incoming governor, Lord De la Warr, just then coming up the river with a fourth resupply. They all turned around and returned to Jamestown, where they slowly began to turn around the colony too. Tough martial-law administration followed under Gates and deputy governor Sir Thomas Dale, and population gradually rose again. But only tobacco assured the economic

The statue of Pocahontas, daughter of Chief Powhatan and supposed savior of Captain John Smith when he was captured by Indians along the Chickahominy River, belongs as much to the mythology as to the history of Virginia. (Colonial National Historical Park)

success that in the eyes of the Company would determine Jamestown's fate. It is a well-known story, how settler John Rolfe, himself a devoted smoker, introduced seed from Trinidad in 1610 or 1611 that developed in Virginia into a hearty new plant: *Nicotiana tabacum.* Early efforts at glass-blowing and silk-making (rats ate the silk worms) would be cheerfully set aside. Shipment of the Weed to England began in 1613 and never ceased. At first it was like a gold rush, and governors had to enforce the production of neglected foodstuffs. Decisively, tobacco thrust settlement outward from Jamestown and became the foundation of colonial Virginia's staple-crop economy.

Rolfe is also remembered for his marriage in 1614 to Chief Powhatan's daughter Pocahontas, a union, we were told in Virginia history class, that strengthened friendly ties with the natives at a crucial period. It did little for Pocahontas, however, who died shortly after Rolfe took her to England. It is also doubtful that relations with the Indians here were much different from what they were everywhere else in America, or not fated to have the same sad outcome. Two races at radically different stages of development, neither very interested in learning about the other except to meet the threat each posed the other, dictated relations that were alternately friendly and hostile. Watchfulness was the only constant, and that could not always prevent disaster, as when in 1622 the Indians massacred 347 of the colony's 1250 inhabitants.

Confronted with a vastly more powerful, more disciplined, and soon to be more numerous culture, the Indians (who were never very many) faded away, their history of it all never written down. For their part, the English who first confronted and subdued them in Virginia were not given to large regret at the fate of, to them, a savage people. Yet there lingers about their encounter an innocent charm reminding us that these lusty Elizabethans (or at least the poets among them) were still given to visions of earthly Edens regained and then as always lost again. This is the context of *The Tempest,* and the theme of Thomas Drayton's more ordinary "Ode to the Virginian Voyage," written at the height of the first Virginia boom in England:

> *You brave heroic minds,*
> *Worthy your country's name,*
>   *That honour still pursue,*
>   *Go and subdue*
> *Whilst loitering hinds*
> *Lurk here at home with shame.*
>
> *Britons, you stay too long,*
> *Quickly aboard bestow you,*
>   *And with a merry gale*
>   *Swell your stretched sail,*
> *With vows as strong*
> *As the winds that blow you . . .*

Captain John Smith, the European name most often associated with the successful founding of Jamestown in Virginia, arrived with the original London Company adventurers in May 1607; it was his soldierly discipline and good relations with the Indians that enabled the fledgling enterprise to survive — barely. (Colonial National Historical Park)

And cheerfully as sea
Success you still entice,
  To get the pearl and gold
  And ours to hold,
Virginia,
Earth's only paradise . . .

Or if you wish, take Captain John Smith's straightforward report of the first English Christmas kept in Virginia: "Among the savages, where we were never more merry, nor fed more plenty of good oysters, fish, flesh, wild fowl and good bread, nor never had better fires in England than in the dry smoky houses of Kecoughtan."

At Jamestown Festival Park elaborate re-enactments of early seventeenth-century life, such as cultivating tobacco, turn the history of Jamestown into summertime theater. (Jamestown-Yorktown Foundation)

Once Jamestown could safely be said to be a going concern, the great age of Atlantic discovery also could safely be said to be over. Not that for years to come there would not be more to find and see in America (in the sense that "no white man had ever seen before"); rather, it was that once the world knew settlement could be accomplished, the point of reference changed. In the age of discovery the most important cargo of those stout little ships had been eastbound news of what was out there. In the age of settlement it was westbound migrants no longer intent on return. Once it was clear that you could find a good life somewhere else, there was (at least not for many generations) no need to go home again.

One of the things, so we were always told, that made a home in the New World so congenial for so many Europeans was that here they might have a voice in the running of their own affairs. Here the language grows slippery—freedom, democracy, representative government—I have confronted them all in great profusion while searching out the history of discovery in America. This is especially so at Jamestown, and such words made up the message that I took away from this place when I first came as a boy.

On July 30, 1619, when the colony had spread up and down the James and when permanence seemed possible, an elective assembly of sorts was convened at Jamestown. As I sit writing with the historical sources for this book arrayed before me, the message pounds home: "This was the beginning of our present system of representative government"; "Here, where the Memorial Church stands, our most cherished traditions of freedom were planted and took root"; "Here the first representative assembly in North America convened, laying the foundations for representative government which we enjoy today"; "A most important development in government took place, from which the ultimate form of American government was shaped . . . A touching scene in its simplicity and yet in all that it signifies—the heart of the political experience of the English speaking peoples and the peculiar contributions they have to make to the world."

These last are the words of scholar A. L. Rowse, written in the

The Jamestown Festival Park has a reconstruction of the glasshouse where the colonists tried unsuccessfully to develop a product saleable in English markets. (Colonial National Historical Park)

1950s and before professional academics had learned to be embarrassed by phrases—and ideas—like "the English speaking peoples." Its usage reaches back as well as ahead in time. Continuity, Rowse is telling us, is at work here, and pretty awesome continuity at that. The author of the first quotation, Charles Hatch, Jr., judiciously continues that the "full intent behind the moves that led to this historic meeting may never be known. It seems to have been another manifestation of the determination to give those Englishmen in America the rights and privileges of Englishmen at home that had been guaranteed to them in the original Company charter. It seems to be this rather than a planned attempt to establish self-government in the New World on a scale that might have been in violation of English law and custom at the time." Then he relents, concluding that whatever the motive, the significance is the same: "This body of duly chosen representatives of the people has continued in existence and its evolution leads directly to our State legislatures and to the Congress of the United States."

If you go, be prepared for this message, with a gloss. You will find the site at the Jamestown National Historic Site, which is jointly administered by the National Park Service and the same Virginia Antiquities ladies who look after the Cape Henry Lighthouse. The Memorial Church, which marks the spot where the assembly first met, stands on the site of the frame church that was built just outside the walls of the fort in 1617. In 1639 it was replaced by a brick structure, and ruins of its bell tower stand here today, the only seventeenth-century structure remaining above ground at Jamestown. The present Memorial Church was a gift of the Antiquities Association in 1907, the tricentenary of the settlement.

We are dealing here, as they want us to believe, with the supposed beginnings of democratic institutions in America, which were not, it is important to remember, unfriendly to the churches. It was pluralist, not secularist, principle that until quite recent times under-lay the idea of separation of church and state. And it was this tension between the notion of a purely lay state (captured in the French *laïque*, which does not translate well into English or American idiom) and an older organic religious sensibility that my companion, and I with him, felt while listening to the ranger's "faith of our fathers" explanation of why these stones command respect.

If you are unlucky and draw a guide other than our old Virginian, and must hear a pale and sanitized version of the story, then stroll over to the memorial pillar nearby, also put up in 1907, and hear without any modern intervention about church and state from the Elizabethans themselves: "Lastly and chiefly," (the advice of the London Council for Virginia to the colonists as they set out in 1606), "the way to prosper and achieve good success is to make yourselves of one mind for the good of your country and your own, and to serve and fear God the giver of all goodness, for every plantation which our heavenly father hath not planted shall be rooted out."

By all of this—the survival through great tribulation, the planting of the seeds of democracy, the piety of a religious people—we are meant to be inspired. Walk the paths past the Tercentenary Monument, the Old Church Tower, the Robert Hunt Shrine, the Dale House, the Memorial Cross, the Third and Fourth State-houses, the statues of Captain John Smith and Pocahontas, and the feeling fills the air. Discovery has become settlement at last, and from this settlement we trace our own sojourn centuries after.

There are, however, two Jamestowns, two separate historic sites pegged to these events. Just across Powhatan Creek from Jamestown National Historic Site sits Jamestown Festival Park, an elaborate historical re-creation, built for the 350th anniversary in 1957 and operated by the Jamestown-Yorktown Foundation, an agency of the Virginia Secretary of Education. This is the place with the three ships I had visited long ago. Wonderfully entertaining as it was then, it is not the sort of place I now care much for, one of a hundred of such reconstructed past worlds, nostalgic fun parks straining hard for modern meanings. Just up the road, colonial Williamsburg sets the standard in the business, with its primly painted "Williamsburg" shops and houses, its costumed interpreters (but oh-so-uncostumed tourists), its manure strewn strangely on black-topped streets, its theme of Becoming American.

Jamestown Festival Park would appear to offer analogous attractions: "Enter the recreated world of seventeenth-century Virginia"

On the tercentenary of the founding of Virginia at Jamestown, commemorators raised an obelisk, which stands today in the Colonial National Historical Park. (Colonial National Historical Park)

The church tower is the only seventeenth-century brickwork remaining above ground at Jamestown. (Colonial National Historical Park)

and you will find an Algonquian Indian village (chief's lodge, family lodge, sweat lodge where medicine men did their business, scarecrow hut where young boys shooed birds off the garden); James Fort (the triangular 1607 palisade with bulwarks, cannon, storehouse, armory, forge and church); and of course the three ships. Using period skills, costumed staff portray the life of the time, demonstrating flintnapping, hide tanning, armor making, military skills and navigational techniques. Presentations, warns the brochure, vary depending on the season.

Good that they do. I wandered here in high summer and pouring rain, and not a lot was going on. The interpreters interpreted precious little, but by their still presence conjured ghosts. Remember: the whole place is a vast cemetery. For years, more died here than lived. Among the uncertain, poorly led living, there were days, weeks and months when not a lot happened either, when the only noises were birds and bugs and the fear in their own heads of savage men and sudden death. A reconstruction, yes, this is—the reconstruction of remoteness.

To walk through the mud, among damp wattle and daub huts, into dark and smoky Indian lodges where "Indian" interpreters split the bones of birds and speak with alarmingly authentic derision of white men's stupidities in the wilderness, is to be yanked back from the accomplishments of settlement to the precariousness of discovery—to be reminded how thin the thread on which it all hung, how slippery the foothold of these seagoing adventurers on a continental land, how likely it all might just as easily have gone the other way. As usual more clear-sighted than most, Captain John Smith well understood the odds and wrote in his famous *Generall Historie of Virginia, New England and the Summer Isles* (Bermuda) that such epic, foolhardy and blood-stirring enterprises as Jamestown "have ever since the world's beginning been subject to such accidents, and everything of worth is found full of difficulties, but nothing so difficult as to establish a commonwealth so far remote from men and means, and where men's minds are so untoward as neither do well themselves nor suffer others."

Epilogue

IT HAS BEEN SAID THAT THE HISTORY OF DISCOVERY belongs more to the sea than to the land, that the accomplishment of the discoverers was chiefly a seaborne accomplishment: epic feats of ocean sailing in tiny ships over great distances and against long odds. All of the discoverers did best at sea. It was for the next generation to bring the same confidence—and luck—to mastering the land. The journeys I have made in order to write this account of the discoverers were coastal journeys, to places where their voyages ended and the utterly different job of settlement began. They were journeys on which, to no surprise, I kept coming upon ships.

The ships (replicas) that can be said to have ended the era of North American discovery—*Susan Constant, Godspeed* and *Discovery*—ride at their moorings at Jamestown in Virginia, where they appear to be a greater attraction to park visitors than any of the exhibits—the fort, the huts, the Indian lodges—that sit on land. The ships were built nearby in 1957 for the occasion of the 350th anniversary of Captain Christopher Newport's landing at Jamestown Island.

The ship that can be said to have begun the era of North American discovery—some Norse *knarr* or longship—sits not at L'Anse aux Meadows in Newfoundland, but oddly enough at the Lincoln Park Zoo in Chicago. Built in Norway, it is a full-size replica of a ninth-century model the likes of which Leif or Thorfinn or Freydis might have sailed in to Vinland long ago. She is called the *Raven* and did in fact make the passage from the Old World to the New early in the summer of 1893 for the occasion of the 400th anniversary of Columbus's discovery of America and the World's Columbian Exposition that gloriously marked the event. I would like to think the *Raven* passed by Quirpon Island and L'Anse aux Meadows en route across the great gulf and up into

Replicated in 1957, Virginia's ships of discovery — Susan Constant, Godspeed and Discovery *— sit at their moorings in the James River, reminders of the colossal accomplishment of discovery. (Jamestown-Yorktown Foundation)*

the St. Lawrence and under the shadows of Stadaconé and Hochelaga and through the chain of Great Lakes and down to the City of Big Shoulders. But it is not so. Sailing from Bergen, Norway, with two mates, eight seamen and a steward, Captain Magnus Anderson took an "all-American" route: forty-four days to New London, Connecticut (and so they did probably sight Newfoundland's Cape Race), then to New York City, up the Hudson River, through the Erie Canal, into the lakes, and down to Chicago, where they arrived for the great exposition on July 12. They did it, certainly, for reasons of Norwegian and American patriotism—to show the flag—and simply to prove that it could be done, reasons that would have made good enough sense to all their, and our, discovery forebears.

The *Raven* followed the Mississippi all the way down the middle of the continent to become, as the Viking Ship Restoration Committee of Evanston, Illinois, tells us, "the first vessel ever to cross the Atlantic from Europe and traverse our entire inland waterway from New York over the Great Lakes to New Orleans." Much of American and Canadian history has to do with subduing vast continental spaces, a job that took, in both cases, the better part of three centuries and that shapes the character of both nations profoundly to this day. Their seaborne heritage from the age of discovery seems remote by comparison. But there is about it something timely as well. To the discoverers, the sea was an obstacle that practice made into a pathway for their successors. Its breadth at first measured the remoteness of the Old World from the New, but in time became the avenue that tied them together. Connectedness is something much on the mind of today's world, where a rejuvenating Europe declares itself once again—as it did in the late Renaissance—to be the shaper of its own destiny. But never again will the Europeans thrust themselves upon the rest of the world (and certainly not the Americas) quite as they did in the age of discovery. Partnership (read as competition among equals) is more today's style, and so Old World and New reach out afresh, across the sea, to shape their future.

The little ships of discovery—*Raven, La Grande Hermine, Susan Constant, Godspeed, Discovery*—also remind us of the connectedness between that time and this. It is a connectedness about which, looking back, there seems such inevitability. At least that is how it is usually rendered and popularly recalled. But look again at the frail hulls and the wildness of the sea and the distance from home, and understand that here is a historical connection truly blessed by fortune, indeed almost providential in nature. Much good, and some evil, came from the North American discovery, and we who live there today are its most immediate beneficiaries. Courage, greed, foolhardiness and faith attended it, qualities that we can say belong as much to our age as to that long-ago adventure. Should we manage as much for our descendants, then our patrimony will have been well repaid.